Compact Textbooks in Mathematics

This textbook series presents concise introductions to current topics in mathematics and mainly addresses advanced undergraduates and master students. The concept is to offer small books covering subject matter equivalent to 2- or 3-hour lectures or seminars which are also suitable for self-study. The books provide students and teachers with new perspectives and novel approaches. They may feature examples and exercises to illustrate key concepts and applications of the theoretical contents. The series also includes textbooks specifically speaking to the needs of students from other disciplines such as physics, computer science, engineering, life sciences, finance.

- **compact:** small books presenting the relevant knowledge
- **learning made easy:** examples and exercises illustrate the application of the contents
- **useful for lecturers:** each title can serve as basis and guideline for a semester course/lecture/seminar of 2-3 hours per week.

Toni Sellarès

Tessellations with Stars and Rosettes

Practical Constructions with Interactive Geometry Software

Toni Sellarès
Departament d'Informàtica, Matemàtica Aplicada i Estadística
University of Girona
Girona, Spain

ISSN 2296-4568　　　　　ISSN 2296-455X　(electronic)
Compact Textbooks in Mathematics
ISBN 978-3-031-82162-2　　　ISBN 978-3-031-82163-9　(eBook)
https://doi.org/10.1007/978-3-031-82163-9

© The Editor(s) (if applicable) and The Author(s), under exclusive license to Springer Nature Switzerland AG 2025

This work is subject to copyright. All rights are solely and exclusively licensed by the Publisher, whether the whole or part of the material is concerned, specifically the rights of translation, reprinting, reuse of illustrations, recitation, broadcasting, reproduction on microfilms or in any other physical way, and transmission or information storage and retrieval, electronic adaptation, computer software, or by similar or dissimilar methodology now known or hereafter developed.
The use of general descriptive names, registered names, trademarks, service marks, etc. in this publication does not imply, even in the absence of a specific statement, that such names are exempt from the relevant protective laws and regulations and therefore free for general use.
The publisher, the authors and the editors are safe to assume that the advice and information in this book are believed to be true and accurate at the date of publication. Neither the publisher nor the authors or the editors give a warranty, expressed or implied, with respect to the material contained herein or for any errors or omissions that may have been made. The publisher remains neutral with regard to jurisdictional claims in published maps and institutional affiliations.

This book is published under the imprint Birkhäuser, www.birkhauser-science.com by the registered company Springer Nature Switzerland AG
The registered company address is: Gewerbestrasse 11, 6330 Cham, Switzerland

If disposing of this product, please recycle the paper.

Competing Interests The author has no competing interests to declare that are relevant to the content of this manuscript.

Contents

1	**Introduction**		1
2	**Preliminaries**		3
	2.1	Isometries	3
	2.2	Tessellations	4
		2.2.1 Periodic Tessellations	4
		2.2.2 Ornamentation with Regular Stars and Rosettes	5
3	**Regular Stars, Rosettes, and Tessellations with Stars and Rosettes**		9
	3.1	Introduction	9
	3.2	Regular Stars	10
		3.2.1 Convergent, Parallel, and Divergent Stars	12
		3.2.2 Construction of Regular Stars	16
		3.2.3 Ornamentation of a Star	18
		3.2.4 Visual Properties of Stars	19
		3.2.5 Coloring of a Star	22
		3.2.6 Ornamentation of the Exterior of a Regular Star	23
	3.3	Rosettes	23
		3.3.1 Convergent, Parallel, and Divergent Rosettes	23
		3.3.2 The Shape of the Petals	23
		3.3.3 Regular, Standard, and Ideal Rosettes	24
		3.3.4 Construction of n-Pointed Rosettes	26
		3.3.5 Visual Properties of Rosettes	29
		3.3.6 Ornamentation of a Rosette	30
		3.3.7 Coloring of a Rosette	30
	3.4	Tessellations with Stars and Rosettes	31
		3.4.1 Determination of the Repeat Unit and, if Necessary, the Base Unit	32
		3.4.2 Construction of Tessellations with Stars and Rosettes Using the Radial Extension Method	32

4 Design of Tessellations with Stars and Rosettes 37
   4.1 Introduction 37
        4.1.1 Charles V Ceiling Room, Royal Alcazar, Seville, Spain 38
        4.1.2 Nasrid Tripod Fountain, Alhambra, Granada, Spain 41
        4.1.3 Tash Hauli Palace Complex, Khiva, Uzbekistan 44
        4.1.4 Court of the Myrtles, Alhambra, Granada, Spain 47
        4.1.5 Nasrid Palace, Alhambra Museum, Granada, Spain 50
        4.1.6 Royal Alcazar, Seville, Spain. 53
        4.1.7 Bibi-Khanum Mosque, Samarkand, Uzbekistan 56
        4.1.8 Shah-i Zinda, Samarkand, Uzbekistan. 59
        4.1.9 Friday Mosque, Isfahan, Iran. 64
        4.1.10 Akbar's Mausoleum, Sikandra, India. 67
        4.1.11 Al-Maridani Mosque, Cairo, Egypt. 71
        4.1.12 Bibi-Khanum Mosque, Samarkand, Uzbekistan 76
        4.1.13 Friday Mosque, Yazd, Iran. 80
        4.1.14 Abdullah Khan Madrasa, Bukhara, Uzbekistan. 84
        4.1.15 Fatima's Haram, Qom, Iran. 87
        4.1.16 Masjid-i-Jami, Varamin, Iran. 91
        4.1.17 Jameh Mosque, Kerman, Iran 94
        4.1.18 Hall of Kings, Alhambra, Granada, Spain. 98
        4.1.19 Bibi Khanum Mosque, Samarkand, Uzbekistan 101
        4.1.20 Qalawun Mausoleum, Cairo, Egypt. 104
        4.1.21 Hot Room, Bath of Comares, Alhambra, Granada, Spain ... 108
        4.1.22 Mashhad al-Imam Yahya ibn al-Qasim, Mosul, Iraq 112
        4.1.23 Tomb of I'timād-ud-Daulah, Agra, India. 116
        4.1.24 Mosaic of the Alhambra Museum, Granada, Spain. 120
        4.1.25 Ulugh Beg Madrasa, Samarkand, Uzbekistan 124
        4.1.26 Ahmad ibn Tulun Mosque, Cairo, Egypt. 128
        4.1.27 Mazar of Sachal Sarmast, Khairpur, Pakistan 131
        4.1.28 Hudavent tomb, Nigde, Turkey 134
        4.1.29 Alhambra, Granada, Spain. 138
        4.1.30 Kok Gumbaz Mosque, Shahrisabz, Uzbekistan. 143
        4.1.31 Alhambra Museum, Granada, Spain 146
        4.1.32 Shah-i Zinda, Samarkanda, Uzbekistan. 149
        4.1.33 Amir Azbek al-Yusufi, Cairo, Egypt 152
        4.1.34 Sabil wa Kuttab al-Sultan Qaytbay, Cairo, Egypt 155
        4.1.35 Ali-Qapu Palace, Isfahan, Iran. 159
        4.1.36 Panel 44 of the Topkapı Scroll, Topkapı Palace museum,
              Istanbul, Turkey 162
        4.1.37 Mosque of Qeycoun, Cairo, Egypt 166
        4.1.38 Masjid Suleyman Pasha, Cairo, Egypt. 169
        4.1.39 Mihrab of the Great Mosque, Damascus, Syria. 172
        4.1.40 Marbel Panel, Great Mosque, Damascus, Syria 175
        4.1.41 Al-Maridani Mosque, Cairo, Egypt. 178
        4.1.42 Gur-e-Amir Mausoleum, Samarkand, Uzbekistan. 183

    4.1.43 Attarine Medersa, Fez, Morocco 187
    4.1.44 Al-Muayyad Mosque, Cairo, Egypt. 191
    4.1.45 Abd al-Ghani al-Fakhri mosque, Cairo, Egypt 195
    4.1.46 Sultan Barsbay Funerary Complex, Cairo, Egypt 199
    4.1.47 Niche in the Comares Palace, Alhambra, Granada, Spain. . . 203
    4.1.48 Two Sisters Room, Alhambra, Granada, Spain 207
    4.1.49 Madrasa al-Bu'inaniya, Fes, Morocco. 213
    4.1.50 Balcony of Lindaraja, Alhambra, Granada, Spain 218

Annotated Bibliography and Webography 223

Further Reading 227

Introduction

Stars and rosettes are among the most characteristic geometric ornamental motifs of Islamic art. They are found on walls, ceilings, doors, and windows in both religious and secular buildings.

Tessellations are repetitive patterns of shapes that fit together without gaps or overlaps. A tessellation is periodic when it can be constructed by translations of a shape, called a repeat unit, in two different directions. Periodic tessellations with stars and rosettes are complex and symmetrical compositions with rhythmic repetitions of profound beauty.

Islamic geometric tessellations can be constructed in several ways: with ruler and compass, with the grid method, with the technique called polygons in contact, or with the modular design system from a small set of basic geometric shapes, among others.

Interactive Geometry Software (IGS), such as GeoGebra, AutoCAD, and Adobe Illustrator, allows you to draw basic geometric elements (points, segments, lines, polygons, vectors, rays, perpendicular and parallel lines, angles, bisectors, perpendicular bisectors, circles, and tangents), calculate angles and distances, and use isometries (translations, rotations, and symmetries) so that you can easily create and then manipulate accurate geometric constructions.

The main objective of this book is to present the radial extension method for generating periodic tessellations with one or more stars or rosettes as central motif and how to use IGS to construct them.

Next, a brief description of the chapters of the book is presented.

Chapter 2 introduces the concepts necessary to understand the contents of the book. Isometries: translation, rotation, central symmetry, and axial symmetry. Periodic tessellation, repeating unit, and base unit.

Chapter 3 formally defines regular stars and rosettes and describes how to construct them accurately using IGS. Next, the radial extension method for generating periodic tessellations with one or more stars or regular rosettes as the central motif using IGS is presented.

Chapter 4 presents 50 models of classic designs of periodic tessellations with stars and rosettes. It describes how to draw each of the models step by step with the radial extension method using IGS.

All the drawings in the book have been made by the author with the free software GeoGebra.

The models presented throughout the book are classic designs of geometric ornamentation in the style of Islamic art, for which this geometric style has an artistic symbology that comes from a philosophy closely linked to religion. Although this philosophical-religious meaning is not discussed in the book, people interested on the topic can refer to the Further Reading at the end of the book.

The prerequisites necessary to understand the book are basic geometric concepts. The book is self-contained, since it provides all the necessary background in Chapter 2.

Overall, the topic of the book is of interest to people attracted to art and geometry because it will help them learn how patterns of profound beauty and intricate designs can be created.

In particular, the book is aimed at students and graduates of mathematics, design, architecture, artists and art historians, as well as anyone who wants to draw tessellations of stars and rosettes using IGS.

The book is suitable for a trimester/semester course/lecture/seminar, with the number of weekly hours depending on the number and difficulty of design models studied. In addition, it can also be used for self-study.

Preliminaries 2

2.1 Isometries

An isometry is a geometric transformation that preserves the distance between pairs of points and, therefore, preserves the lengths, angles, and areas of geometric objects, and consequently also their shape and size.

Isometries include translation, rotation, central symmetry, and axial symmetry (see Fig. 2.1).

- Translation: Moves a point a fixed distance in a given direction and sense (translation vector t).
- Rotation: Circularly moves a point around a fixed point O (center of rotation) by a given angle α (angle of rotation).
- Central symmetry: Moves a point with respect to a point O (center of symmetry) so that the middle point of the segment that joins the initial point and the moved point is the center of symmetry. It is equivalent to a rotation of 180° of center O.
- Axial symmetry: Moves a point with respect to a line s (axis of symmetry) so that the bisector of the segment that joins the initial point and the moved point is the axis of symmetry.

An isometry of a geometric object is obtained by applying the isometry to each of its points.

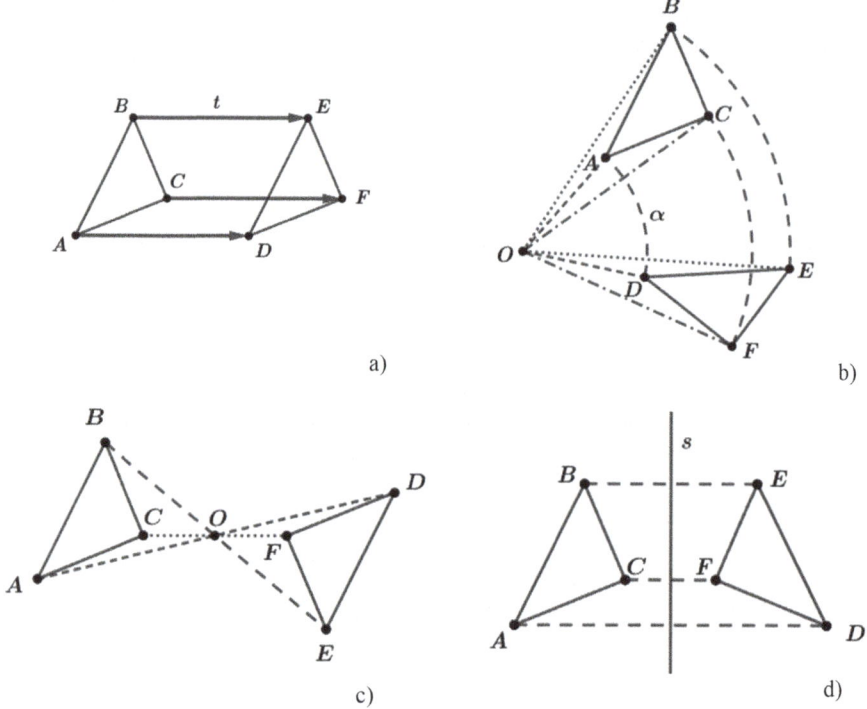

Fig. 2.1 Isometries in which points A, B, and C become points D, E, and F: (**a**) Translation; (**b**) Rotation; (**c**) Central symmetry; (**d**) Axial symmetry

2.2 Tessellations

A tessellation is a covering of a flat surface with closed figures, shapes, so that there are no overlaps or empty spaces.

A tessellation is polygonal if the shapes are polygons. A polygonal tessellation is side-by-side if all polygons that are adjacent share a full side.

A tessellation that uses only one type of shape is said to be monohedral.

2.2.1 Periodic Tessellations

A tessellation is periodic when it can be constructed by translations in two different directions of a set formed by one or more adjacent shapes, called the repeat unit (Fig. 2.2).

A base unit is a shape that generates the repeat unit through rotations or axial symmetries or a combination of both (Fig. 2.3).

2.2 Tessellations

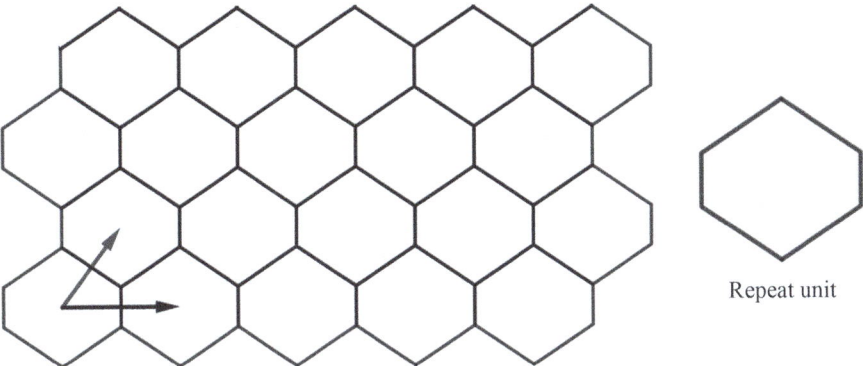

Fig. 2.2 Polygonal side-by-side monohedral periodic tessellation

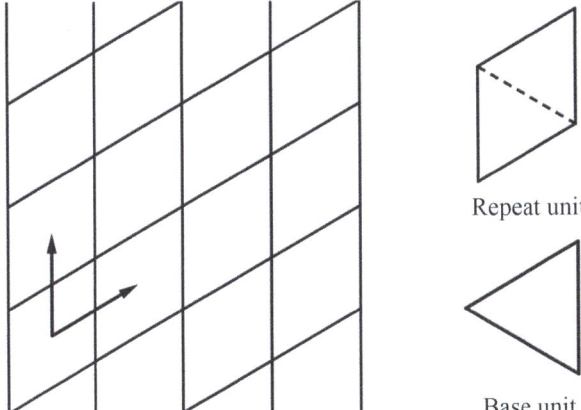

Fig. 2.3 Polygonal side-by-side monohedral periodic tessellation

2.2.2 Ornamentation with Regular Stars and Rosettes

In many side-by-side polygonal monohedral periodic tessellations that have geometric ornamentation in the style of Islamic art, the repeat unit is centrally decorated with one or more regular stars or rosettes, which are called central motifs.

When the repeat unit contains more than one central motif, it is constructed from a shape, the base unit, which contains a single central motif and populates the repeating unit by means of axial symmetries, central symmetries, or rotations.

The part of the repeat or base unit outside the central motif is filled so that the parts bordering their adjacent ones in the final tessellation match correctly.

Figures 2.4, 2.5, 2.6, and 2.7 show examples in which the repeat or the base unit is centrally ornamented with a regular star or rosette.

Tessellation that can be found in the Friday Mosque, Isfahan, Iran

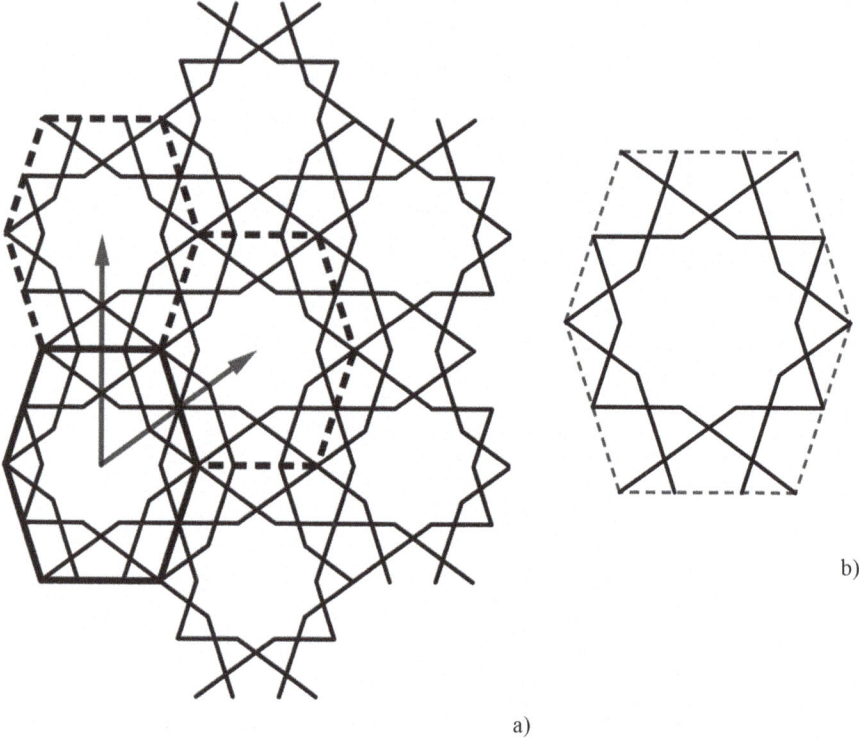

Fig. 2.4 (a) Polygonal side-by-side monohedral periodic tessellation ornamented with regular stars; (b) Repeat unit decorated with a regular star

2.2 Tessellations

Tessellation that can be found in a mosaic in the Alhambra Museum, Granada, Spain

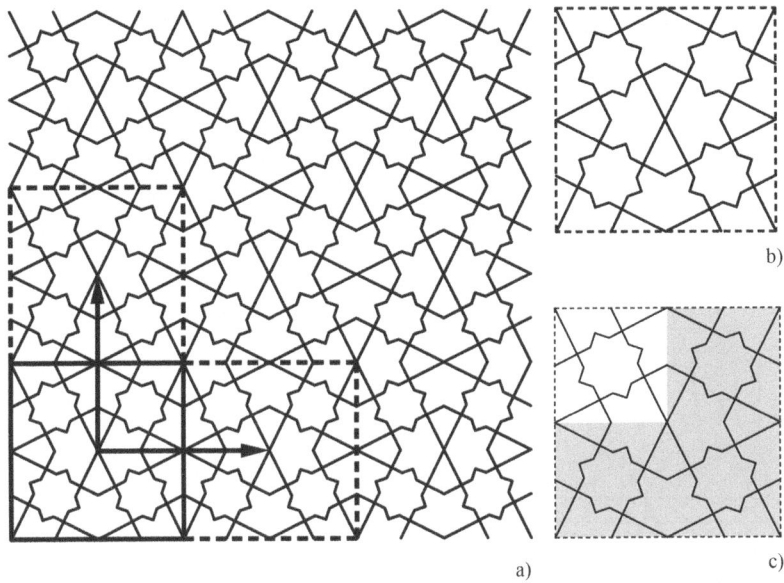

Fig. 2.5 (**a**) Polygonal side-by-side monohedral periodic tessellation ornamented with regular stars; (**b**) Repeat unit decorated with four regular stars; (**c**) Base unit decorated with a regular star

Tessellation that can be found in Masjid Suleyman Pasha, Cairo, Egypt

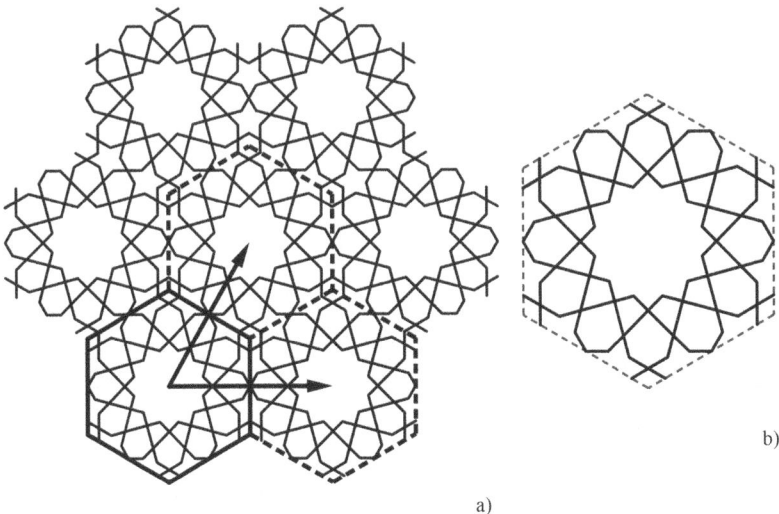

Fig. 2.6 (**a**) Polygonal side-by-side monohedral periodic tessellation ornamented with rosettes; (**b**) Repeat unit decorated with a rosette

Tessellation that can be found in the Balcony of Lindaraja, Alhambra, Granada, Spain

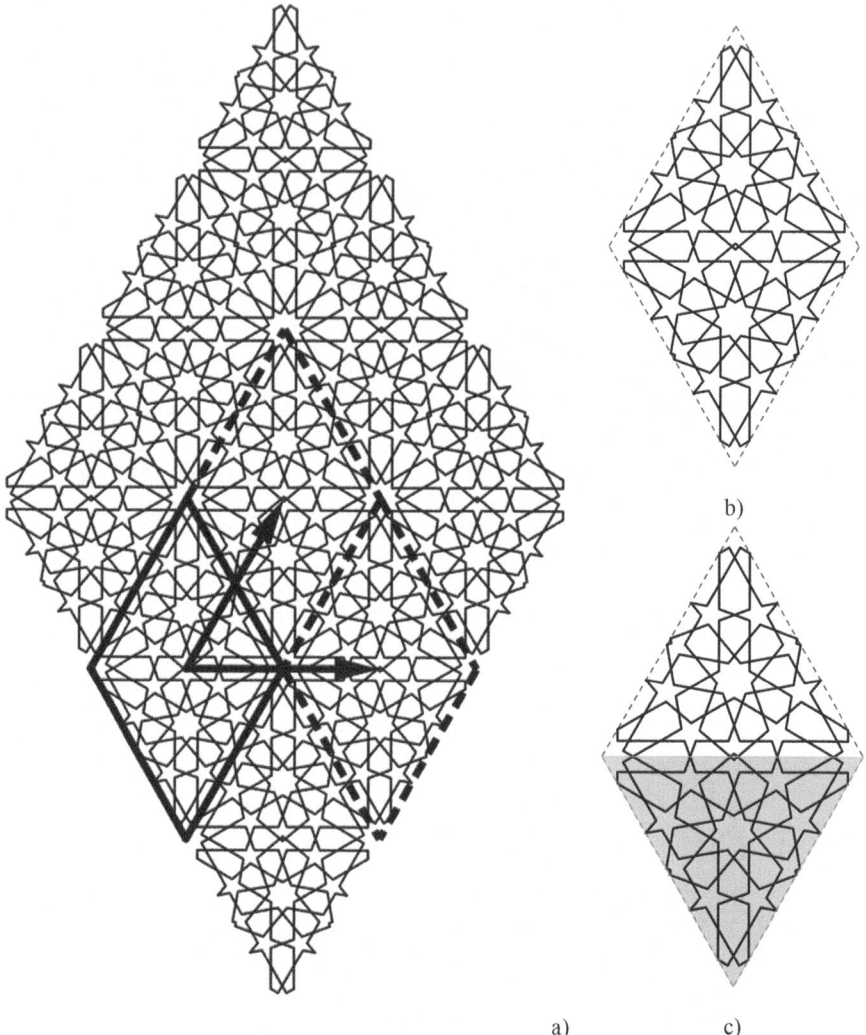

Fig. 2.7 (**a**) Polygonal side-by-side monohedral periodic tessellation ornamented with rosettes; (**b**) Repeat unit decorated with two rosettes; (**c**) Base unit decorated with a rosette

Regular Stars, Rosettes, and Tessellations with Stars and Rosettes

3.1 Introduction

Stars and rosettes are among the most characteristic ornamental motifs of Islamic art (Fig. 3.1).

With stars and rosettes, complex and symmetrical tessellations can be constructed with rhythmic repetitions of profound beauty. We find them on walls, ceilings, doors, and windows, both in religious and secular buildings (Fig. 3.2).

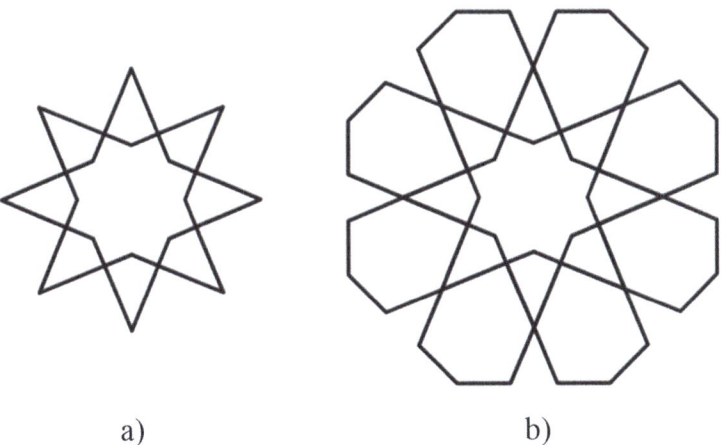

a)   b)

Fig. 3.1 Examples of: (**a**) regular star; (**b**) rosette

Fig. 3.2 Tessellation: (**a**) with stars. Friday Mosque, Isfahan, Iran; (**b**) with rosettes. Balcony of Lindaraja, Granada, Spain

3.2 Regular Stars

A *regular n-pointed star*, n ≥ 3, is a non-convex polygon that has (see Fig. 3.3):

(a) 2n sides of equal length
(b) Two sets of n vertices that alternate on two concentric circles. The n vertices of the same type are separated by arcs of angle α = 360°/n and adjacent vertices of different type are separated by arcs of angle α/2 = 180°/n

The center of the regular star is the center O of the concentric circles. A spike is formed by the two sides adjacent to an outer vertex of the star. A dent is formed by the two sides adjacent to an inner vertex of the star. All spike vertices have angle β and all the dent vertices have outer angle ϒ.

The number n and the angle β uniquely determine a regular n-pointed star, which we will represent by (n, β).

We will have (see Fig. 3.4):

The angle: **α = 360°/n**.
The angles of the triangle OP'R are equal to α/2, 90°—α/2 and 90°.

The angle OP'Q must be smaller than the angle OP'R, from which:

$$\beta/2 < 90° - \alpha/2; \beta < 180° - \alpha; \beta < 180°(1 - 2/n).$$

The angles of the triangle OPQ' are equal to α/2, β/2 and 180°-ϒ/2, therefore:

$$\alpha/2 + \beta/2 + (180° - \Upsilon/2) = 180; \Upsilon = \alpha + \beta.$$

There is a second equivalent way of determining an n-pointed regular star, which we describe below (see Fig. 3.5).

Let: O be the center of the star; P an outer vertex and Q its adjacent inner vertex clockwise; s the line that passes through P and Q; D the point of intersection of the

3.2 Regular Stars

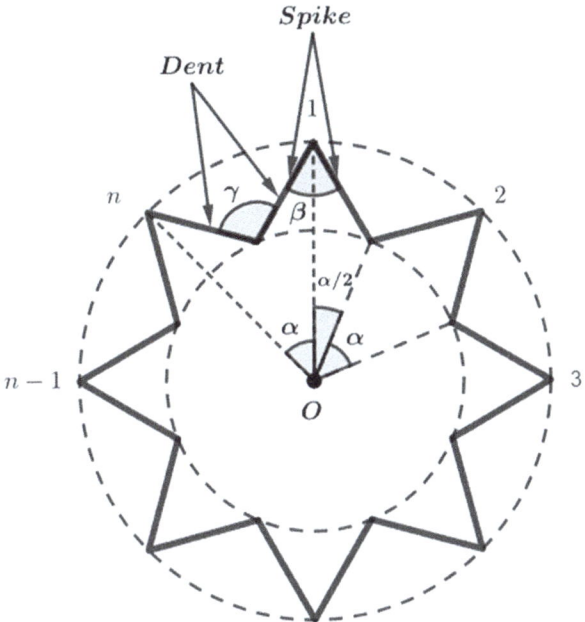

Fig. 3.3 Regular star (n, β)

Fig. 3.4 Properties of the angles of regular stars

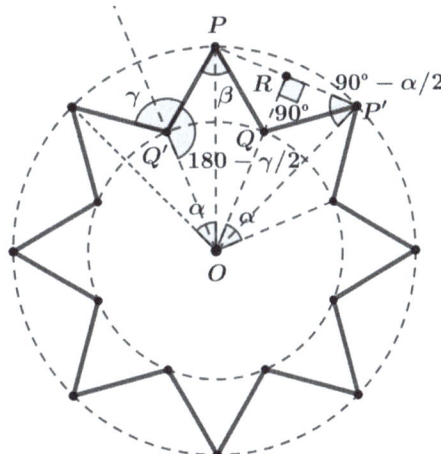

line s with the outer circumference and δ the angle formed by the segments OP and OD.

We define $\mathbf{q = \delta/\alpha}$.

From the isosceles triangle ODP, we deduce that:

$$\beta/2 + \beta/2 + \delta = 180°, \delta = \mathbf{180° - \beta},$$

Fig. 3.5 An equivalent way of determining an n-pointed regular star

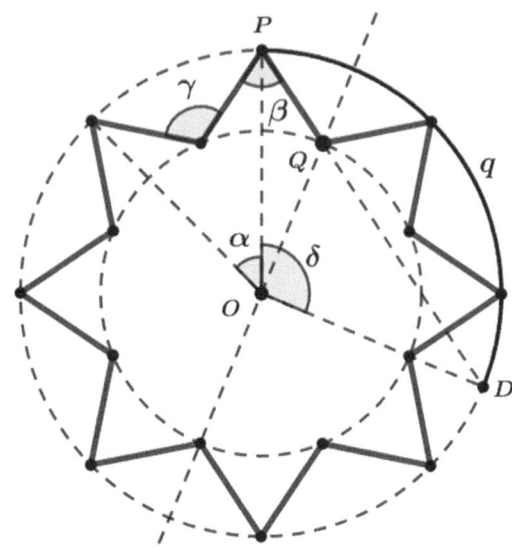

thus:

$$q = \delta/\alpha = (180° - \beta)/\alpha = 180°/\alpha - \beta/\alpha = (n/2)(1 - \beta/180°)(*),$$

from which it follows:

$$\beta = 180°(1 - 2q/n)(**).$$

Thus, knowing n and β we can find q and knowing n and q we can find β. Therefore, the regular star (n, β) is also uniquely determined by n and q. When we use the parameters n and q to define a regular star, we will denote it by the expression |n/q|. When β and q satisfy (*) and (**) we have: **(n, β) = |n/q|**.

Figure 3.6 shows examples of regular stars (8.45°) = |8/3| and (8.67.5°) = |8/2.5|.

When β tends to 180° - α then q tends to 1 and vice versa. In this case, the star has wide and flat spikes (Fig. 3.7).

Note that, when β = 180° - α = 180–360°/n we have a regular polygon with n sides.

When β tends to 0° then q tends to n/2 and vice versa. In this case, the star has narrow and sharp spikes (Fig. 3.8).

From the previous observation it follows that, for any value of n, it is always **q < n/2**.

3.2.1 Convergent, Parallel, and Divergent Stars

Let s and t be the outward extension of the two sides furthest from two adjacent points of a star.

We will use that:

3.2 Regular Stars

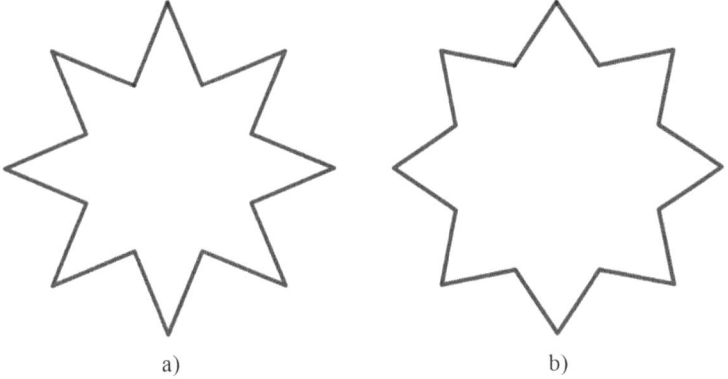

Fig. 3.6 (a) Regular star (8.45°) = |8/3|; (b) Regular star (8.67.5°) = |8/2.5|

Fig. 3.7 β close to 180° - α and q close to 1

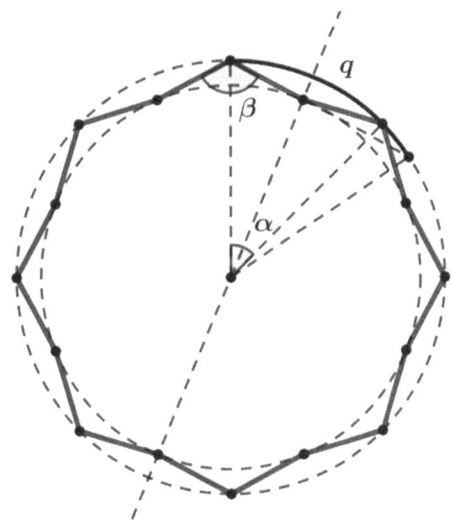

$$\Upsilon = \alpha + \beta;$$
$$\mathbf{q} = \delta / \alpha = (180° - \beta) / \alpha = 180° / \alpha - \beta / \alpha = 180° / (360° / n) - \beta / \alpha = \mathbf{n/2} - \beta / \alpha$$

The rays s and t converge if and only if $\beta > \Upsilon/2$ and then we say that the *star is convergent* (Fig. 3.9).

$$\text{If } \beta > \Upsilon / \mathbf{2} = \alpha/2 + \beta/2 \text{ then } \alpha < \beta$$

and

$$\text{from } \beta / \alpha > 1 \text{ results } \mathbf{q} = n/2 - \beta/\alpha < \mathbf{n/2 - 1}$$

The rays s and t are parallel if and only if $\beta = \Upsilon/2$ and then we say that the *star is parallel* (Fig. 3.10).

Fig. 3.8 β close to 0° and q close to n/2

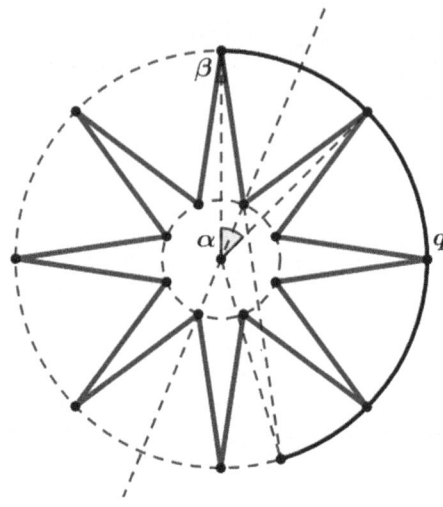

If $\beta = \Upsilon/2 = \alpha/2 + \beta/2$ then $\alpha = \beta$

Fig. 3.9 Convergent star: $\alpha < \beta$ and $q < n/2-1$

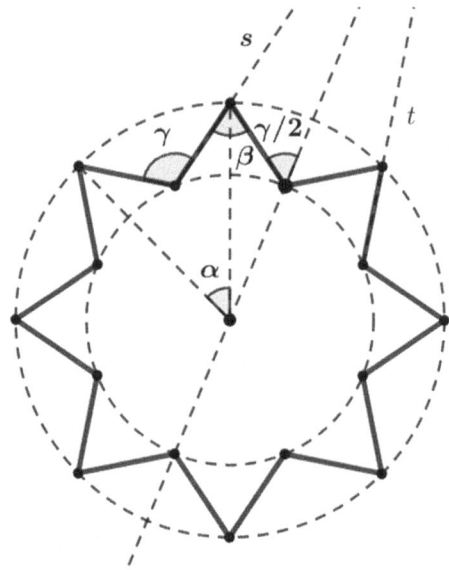

and

from $\beta/\alpha = 1$ it results $q = n/2 - \beta/\alpha = \mathbf{n/2 - 1}$

The rays s and t diverge if and only if $\beta < \Upsilon/2$ and then we say that the *star is divergent* (Fig. 3.11).

3.2 Regular Stars

Fig. 3.10 Parallel star: $\alpha = \beta$ and $q = n/2-1$

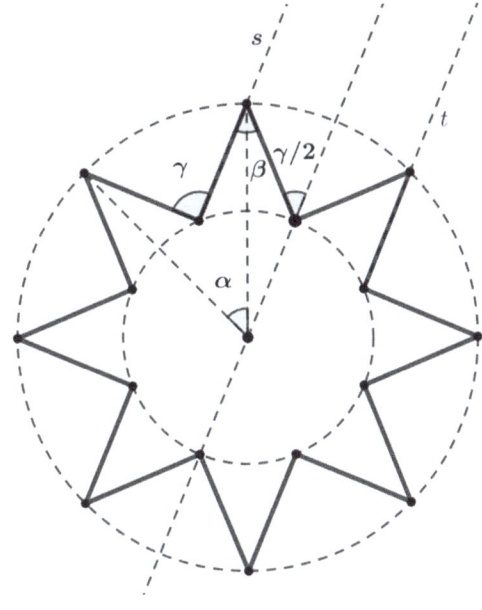

Fig. 3.11 Divergent star: $\alpha > \beta$ and $q > n/2-1$

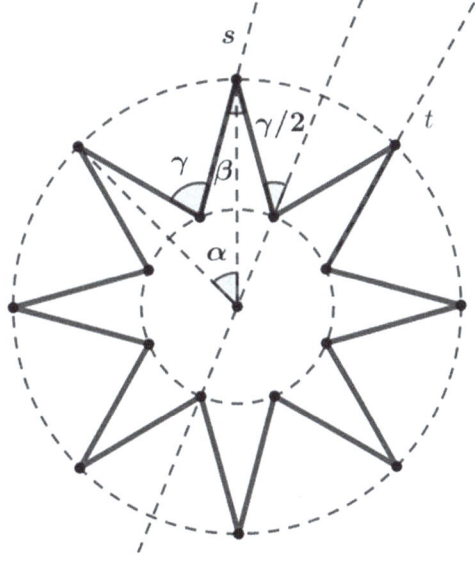

If $\beta \langle \Upsilon/2 = \alpha/2 + \beta/2$ then $\alpha \rangle \beta$

and

of $\beta/\alpha \langle 1$ results $\mathbf{q} = n/2 - \beta/\alpha \rangle \mathbf{n/2-1}$

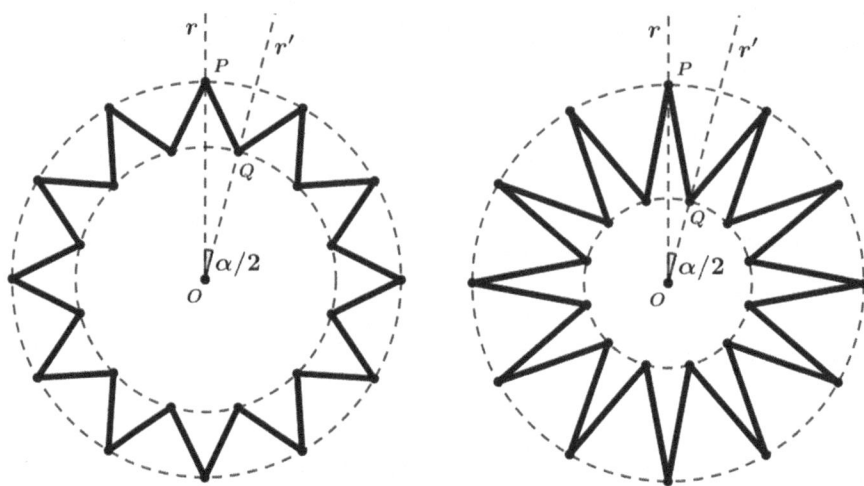

Fig. 3.12 Examples of 12-pointed stars obtained by varying the position of point Q on the ray r′

3.2.2 Construction of Regular Stars

To construct a regular n-pointed star (see Fig. 3.12), first we draw the ray r of origin O that passes through an exterior vertex P and the ray r′ that is obtained from r by a rotation of center O and angle $\alpha/2 = 180°/n$ clockwise. As vertex adjacent to point P clockwise, we can choose any point Q of r′ such that the distance from O to Q is less than the distance from O to P. Then we complete the two sets of vertices that define the regular star by successive rotations of P and Q of center O and angle α clockwise. The n exterior vertices belong to a circle and the other n vertices to a second circle inside the first. Finally, we build the star by drawing the polygon that alternately joins the points of the outer and inner circles that define the regular star. By varying the position of point Q on the ray r′ we obtain different regular stars of n points.

If we want to draw a regular star (n,β) or |n/q|, then the position of point Q is predetermined. Whether we know β or we know q, first we need to find the ray s that will contain the side PQ of the star and then calculate Q as the intersection of the rays r′ and s. What changes is the way to find the ray s (see Fig. 3.13):

– Given β:

Draw: the ray s rotation of the ray r with a rotation of center P and angle β/2 counterclockwise.

– Given q:

3.2 Regular Stars

Fig. 3.13 Determination of the ray s

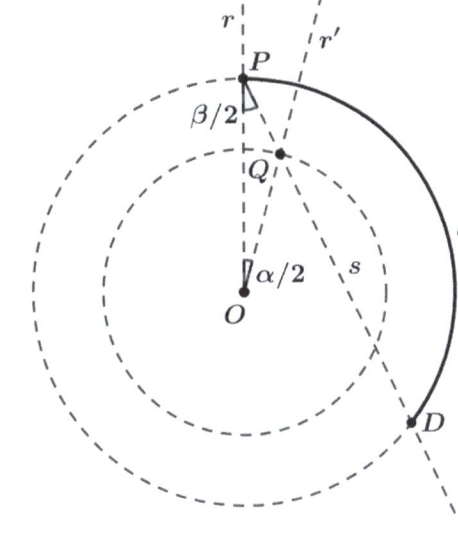

Fig. 3.14 Regular star $(5.36°) = |5/2|$

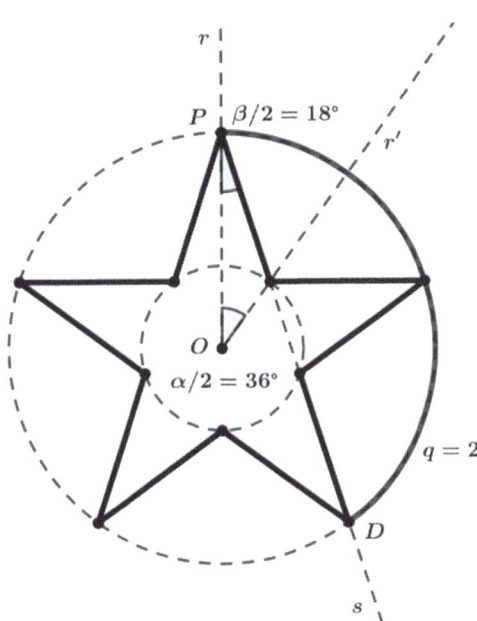

Draw: the point D rotation of the point P of center O an angle $\alpha = 360° \, q/n$ anticlockwise; the ray s of origin P that passes through D.

Knowing the point Q, we proceed as we explained before to draw the regular star (n,β) or $|n/q|$.

Figures 3.14 and 3.15 show examples of regular stars.

Fig. 3.15 Regular star (9.40°) = |9/3.5|

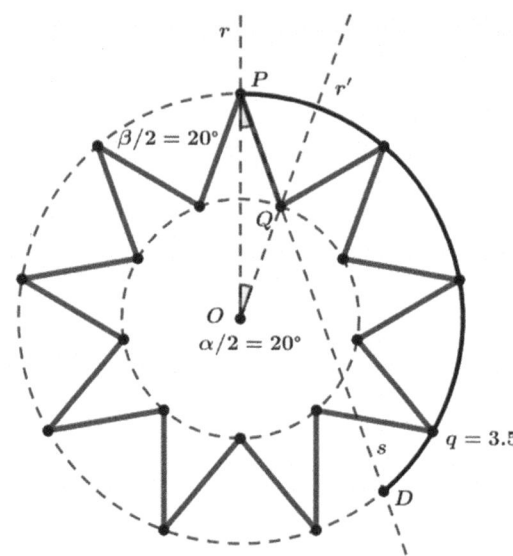

3.2.3 Ornamentation of a Star

It is customary to decorate the empty space in the central part of a regular star. The following describes the process of adding a layer to a regular n-pointed star to generate a new central star, also n-pointed, surrounded by n kite-shaped quadrilaterals. The resulting ornate star has two layers (see Figs. 3.16, 3.17 and 3.18).

1. Draw the point S intersection of the rays of origin the vertex P′ passing through the point Q and the ray of origin the vertex P″ passing through the point R
2. Draw the missing vertices of the new central star by successive rotations of S of center O and angle 360°/n clockwise. Complete the drawing as shown

From the angles of the triangle in the following figure, we deduce (Fig. 3.19):

$$\alpha/2 + \beta/2 + \beta'/2 + 180° - \Upsilon = 180°$$

from where: **β' = β + α.**
If even more inner layers are drawn, then the number of layers c must meet:

$$\beta + (c-1)\alpha < 180 - \alpha$$

from where: **c < (180 - β)/α = n(180 - β)/360 = q**

Therefore, since always q < n/2, also c < n/2. Consequently, no more than one layer can be drawn if n < 5.

When a regular star is represented with c layers, we denote it (n, β)c = |n/q|c.
Figure 3.20 shows two regular stars with the maximum number of layers. Usually only 2 layers are drawn (see Fig. 3.21).

3.2 Regular Stars

Fig. 3.16 Ornamentation of a star. Step 1

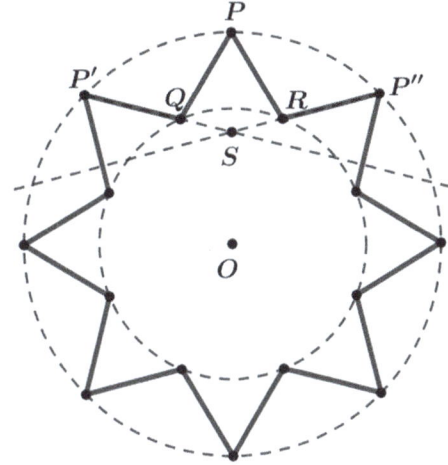

Fig. 3.17 Ornamentation of a star. Step 2

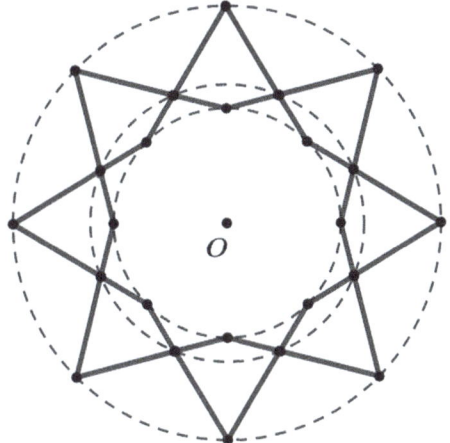

3.2.4 Visual Properties of Stars

The pair of stars in Fig. 3.22, although geometrically identical, has different visual properties. Figure 3.22a, in which the vertical and horizontal directions dominate, convey the feeling of balance and stability, while Fig. 3.22b, in which the diagonal directions dominate, convey the feeling of dynamism and movement.

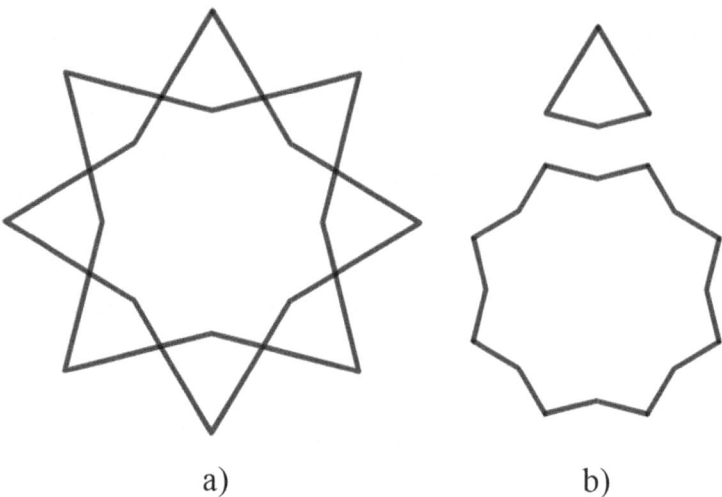

Fig. 3.18 (a) Star with two layers; (b) Kite and central star

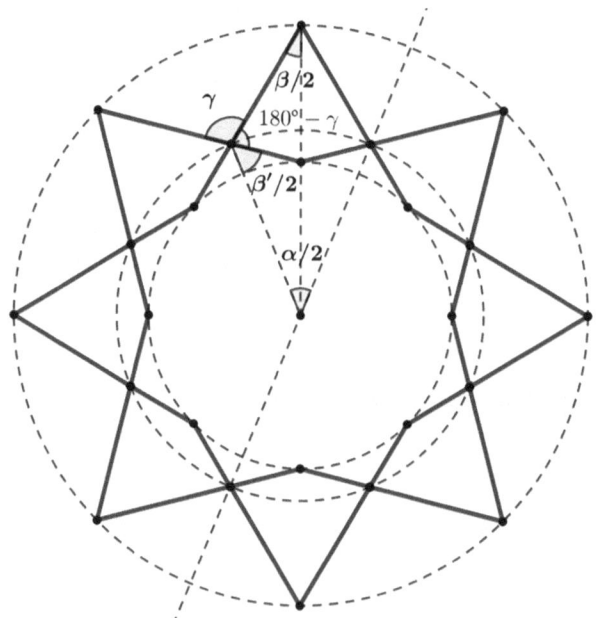

Fig. 3.19 Relation between the angles β and β' of successive layers

3.2 Regular Stars

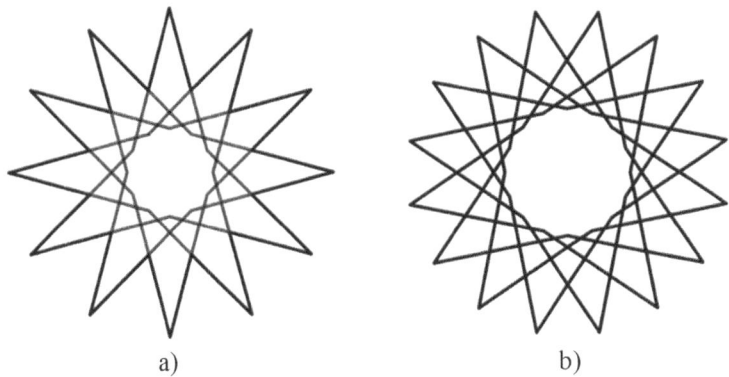

Fig. 3.20 (a) Regular star (12, 30°)4 = |12/5|4; (b) Regular star (16, 45°)5 = |16/6|5

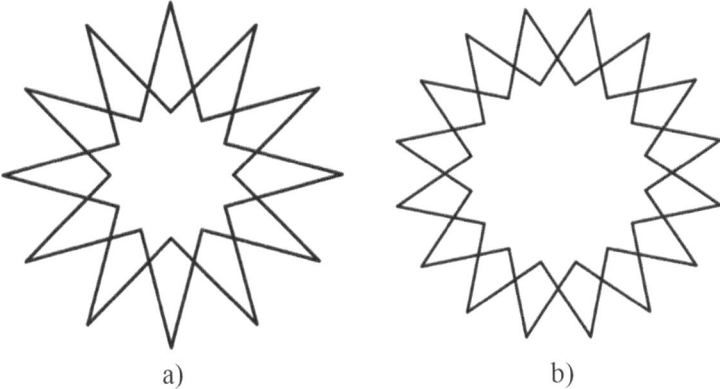

Fig. 3.21 (a) Regular star (12, 30°)2 = |12/5|2; (b) Regular star (16, 45°)2 = |16/6|2

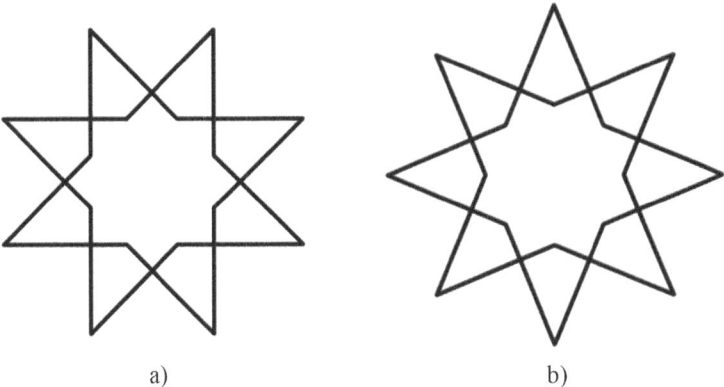

Fig. 3.22 Different visual properties depending on the dominant directions

3.2.5 Coloring of a Star

The same star can have different aesthetic qualities depending on whether it is represented by lines or in the form of a mosaic coloring the different layers (see Figs. 3.23 and 3.24).

By doing the opposite process, one can add a layer to the outside of a star of angle β' and obtain a new star of angle $\beta = \beta' - \alpha$, with the condition that $\beta' - \alpha > 0$ and therefore $\beta' > \alpha$.

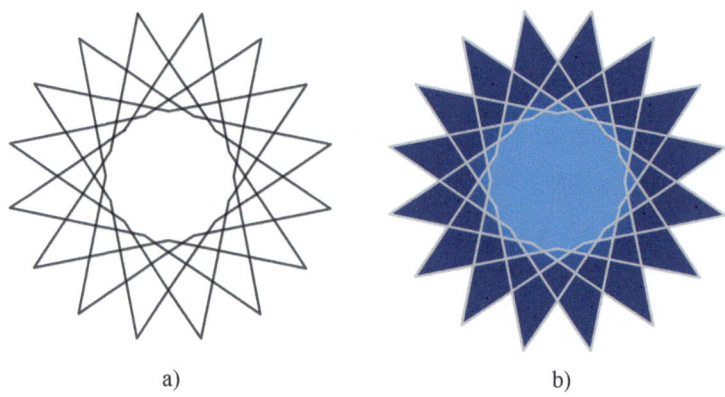

Fig. 3.23 Regular star |16/6|5: (**a**) represented by lines; (**b**) in the form of colored mosaic

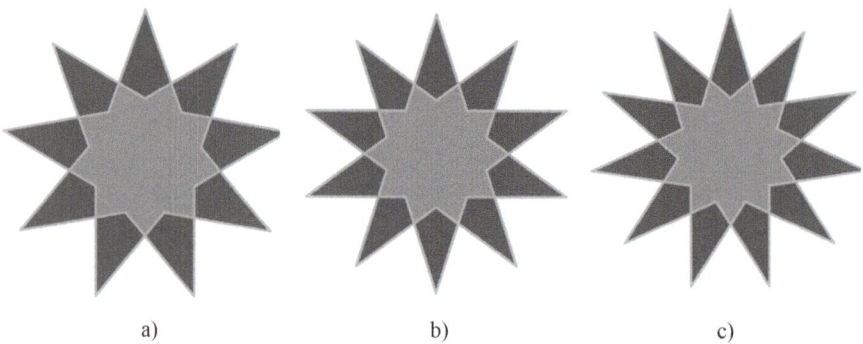

Fig. 3.24 Colored parallel regular stars: (**a**) $(9,40°)2 = |9/3.5|2$; (**b**) $(10,36°)2 = |10/4|2$; (**c**) $(11,32.727°)2 = |11/4.5|2$

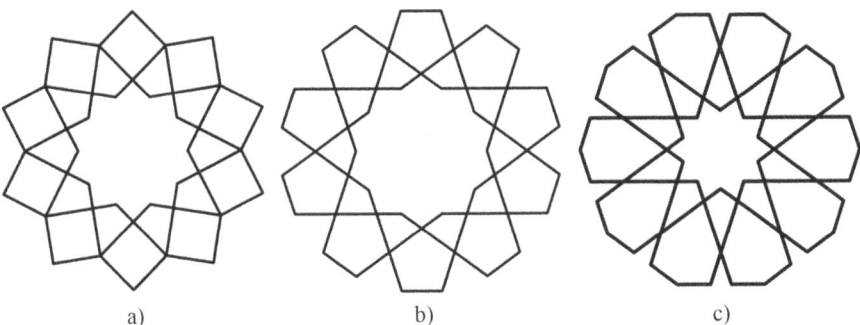

Fig. 3.25 Exterior ornamentation of a regular star with: (**a**) squares; (**b**) pentagons; (**c**) hexagons

3.2.6 Ornamentation of the Exterior of a Regular Star

The exterior of a regular star may also be adorned, for example, with a ring of squares, pentagons, or hexagons (Fig. 3.25).

3.3 Rosettes

A *n-pointed rosette* is formed by a regular star $(n,\beta)c = |n/q|c$ surrounded by n identical hexagons called *petals*. Usually c = 1, 2 or 3, but the most common value is 2 (see Fig. 3.26).

A petal fits between the two sides of a dent of the regular star.

A petal is symmetrical about the line that passes through its outermost vertex and the center of the star.

3.3.1 Convergent, Parallel, and Divergent Rosettes

From a base star $(n,\beta)c = |n/q|c$ which is convergent ($360/n < \beta$ and $q < n/2-1$), parallel ($360/n = \beta$ and $q = n/2-1$), or divergent ($360/n > \beta$ and $q > n/2-1$), a n-pointed rosette is generated such that when extended outwards the two lateral sides of any petal converge, are parallel or diverge and, consequently, we will say that the rosette is convergent, parallel, or divergent (Fig. 3.27).

Parallel and convergent rosettes are the most common in Islamic art.

3.3.2 The Shape of the Petals

The length of the petals is not determined by the geometry of the base star and depends on the designer's taste or the needs to fit when the rosette is part of a larger composition.

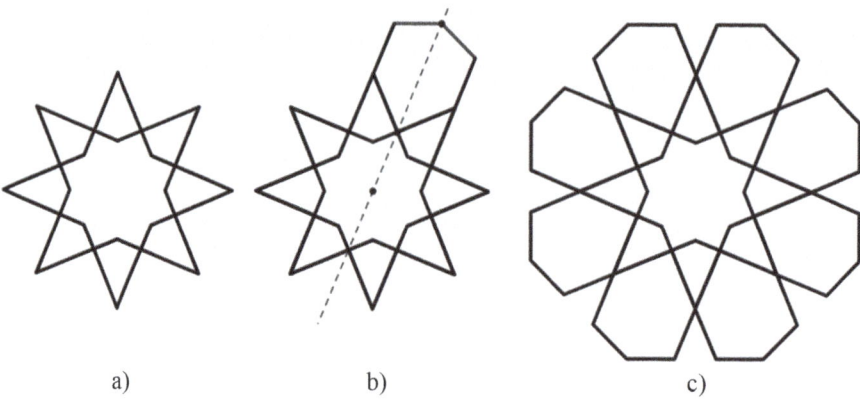

Fig. 3.26 (**a**) Star |8/3|2; (**b**) Fit of a petal between two adjacent sides of the star and axial symmetry of the petal; (**c**) Rosette of 8 petals

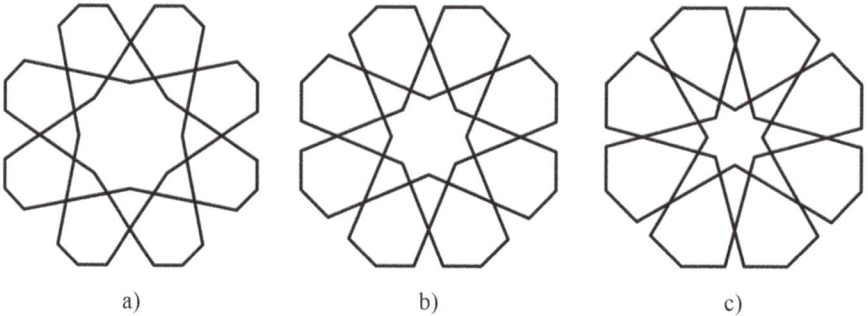

Fig. 3.27 Rosettes: (**a**) convergent; (**b**) parallel; (**c**) divergent

Longer petals are usually used for high n values, so that the tips of the petals are far enough apart to be more easily visible (Fig. 3.28).

3.3.3 Regular, Standard, and Ideal Rosettes

A n-pointed rosette is said to be regular if the two sides adjacent to the vertex of the petals are part of the outline of a regular polygon with n sides concentric with the rosette (Fig. 3.29).

Since the angle φ of the vertex of a petal of a regular rosette coincides with the interior angle of the regular polygon that delimits it, we have: φ = 180–360°/n.

A rosette is said to be standard if the four outer sides of a petal are of the same length (Fig. 3.30).

A rosette is said to be ideal if it is both regular and standard (Fig. 3.31).

Since an ideal rosette is regular, the angle φ of the tip of a petal is: φ = 180–360°/n.

3.3 Rosettes

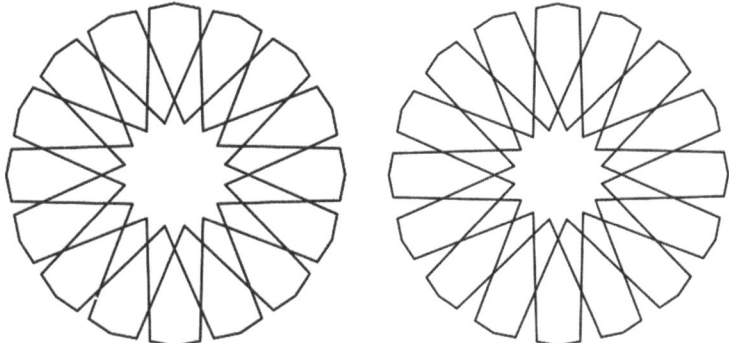

Fig. 3.28 Rosettes with different petal lengths

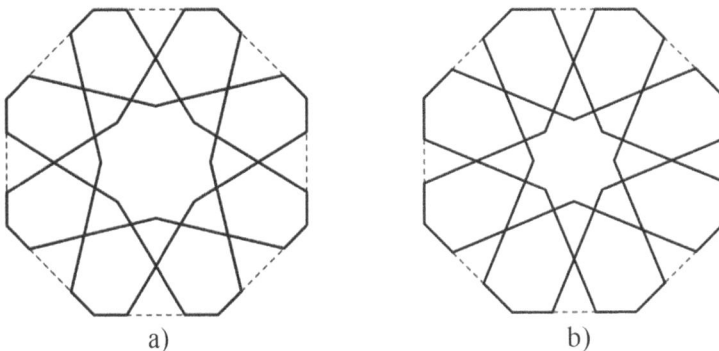

Fig. 3.29 Regular rosettes: (**a**) convergent; (**b**) parallel

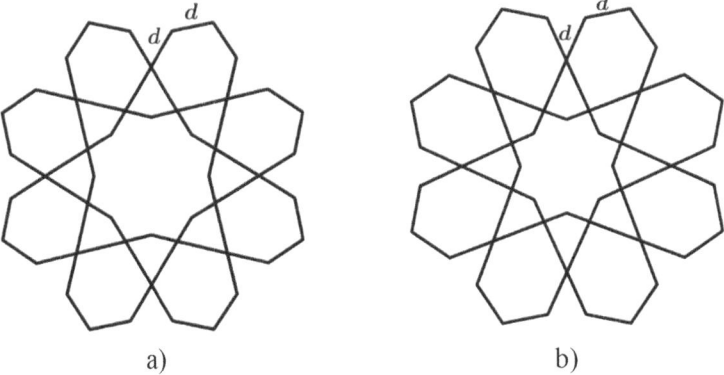

Fig. 3.30 Standard rosettes: (**a**) convergent; (**b**) parallel

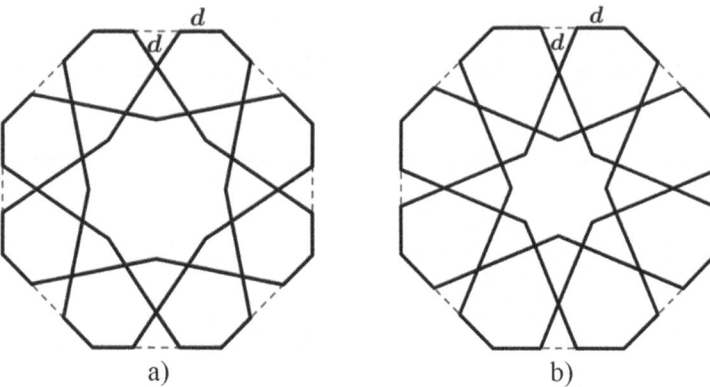

Fig. 3.31 Ideal rosettes: (**a**) convergent; (**b**) parallel

Fig. 3.32 Initial regular star of n points

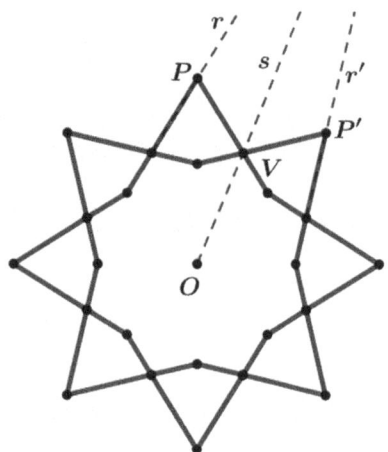

3.3.4 Construction of n-Pointed Rosettes

Start from an n-pointed regular star (see Fig. 3.32). Draw: the outward extensions r and r' of the external sides of two adjacent exteriors vertices P and P' of the star; the ray s of origin the center O of the star and passing through the common interior vertex V at both ends.

3.3.4.1 General Construction

The most general way to represent a petal of a rosette that fits between the two sides of a dent of a regular star consists of (see Fig. 3.33):

Draw: a point L on r as the outer lateral vertex of the petal; the point L' of r' symmetric of L with respect to s; a point E on s as the tip of the petal.

The rosette is finished drawing the set of n petals by successive rotations, with center O and angle $\alpha = 360°/n$ counterclockwise, of the petal initially represented.

Fig. 3.33 General construction of a rosette petal

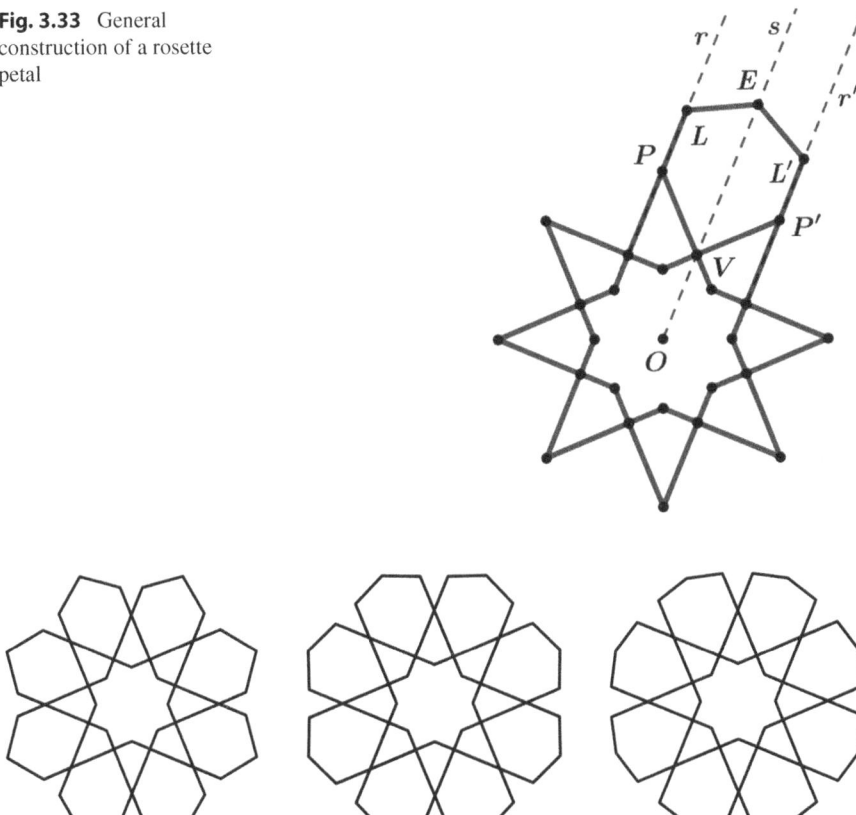

Fig. 3.34 Parallel rosettes obtained by varying points L and E

By varying the point L on r and the point E on s we obtain different rosettes (Fig. 3.34).

3.3.4.2 Construction of Regular Rosettes

Next, we describe how to draw the three outer vertices of a petal of a regular rosette (see Fig. 3.35).

Draw: a point E on s which will be the tip of the petal; the tip E' of the petal adjacent to that of tip E obtained by rotating E with a rotation of center O and angle $\alpha = 360°/n$ counterclockwise; the line l through E and E'; the vertex L of the petal intersection of r and l; the vertex L' on r' symmetric of L with respect to s. The points L, E, L' are the three outer vertices of the petal. Note that the ray of origin O through P and the line l are perpendicular.

Finish building the rosette by completing the set of n petals by successive turns, with center O and angle $\alpha = 360°/n$ counterclockwise, of the petal initially represented.

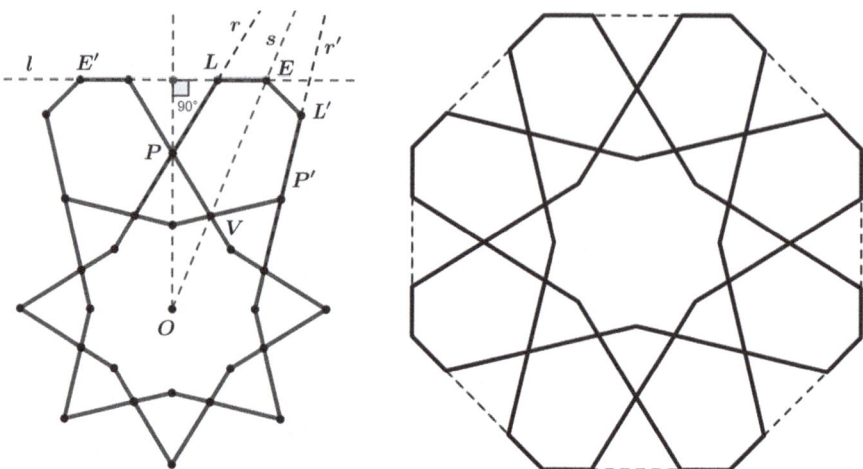

Fig. 3.35 Construction of a regular rosette (non-standard)

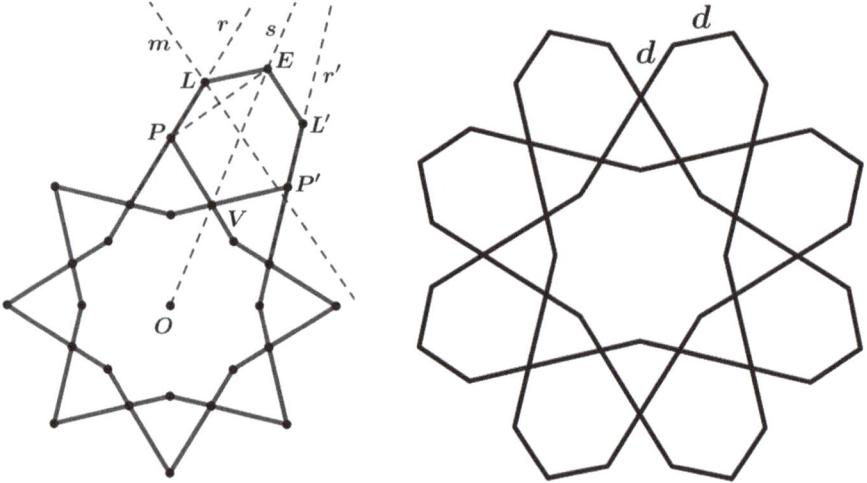

Fig. 3.36 Construction of a standard rosette (non-regular)

By varying the point E on s we obtain different regular rosettes.

3.3.4.3 Construction of Standard Rosettes

Below we describe how to build a petal of a standard rosette (Fig. 3.36).

Draw: a point E on s which will be the tip of the petal; the perpendicular bisector m of the segment with endpoints the vertex P of the star and E; the vertex L of the petal intersection of r and m; the point L' of r' symmetric of L with respect to s. Points L, E, L' are the three outer vertices of the petal.

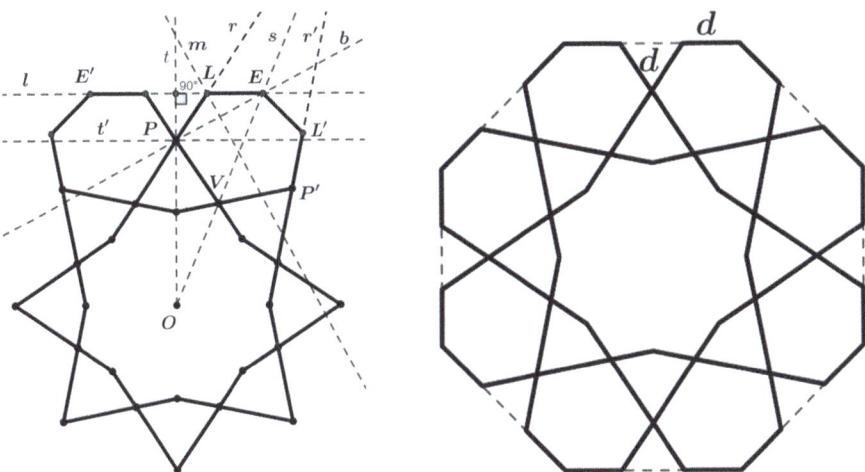

Fig. 3.37 Construction of an ideal rosette

The rosette is finished completing the set of n petals by successive rotations, of center O and angle α = 360°/n counterclockwise, of the petal initially represented.

By varying the point E on s we obtain different standard rosettes.

3.3.4.4 Construction of Ideal Rosettes

We describe how to build a petal of an ideal rosette (Fig. 3.37). Draw: the ray t of origin the center O of the star passing through the point P; the perpendicular t' to t passing through P; the bisector b of r and t'; the tip E of the petal intersection of b and s; the perpendicular bisector m of the segment with endpoints P and E; the vertex L of the petal intersection of r and m; the point L' of r' symmetric of L with respect to s. The points L, E, L' are the three outer vertices of the petal.

We finish building the rosette by completing the set of n petals by successive rotations, of center O and angle α = 360°/n counterclockwise, of the petal initially represented.

3.3.5 Visual Properties of Rosettes

The pair of rosettes in the figures below, although geometrically identical, has different visual properties. Figure 3.38a, in which the vertical and horizontal directions dominate, convey the feeling of balance and stability, while Fig. 3.38b, in which the diagonal directions dominate, convey the feeling of dynamism and movement.

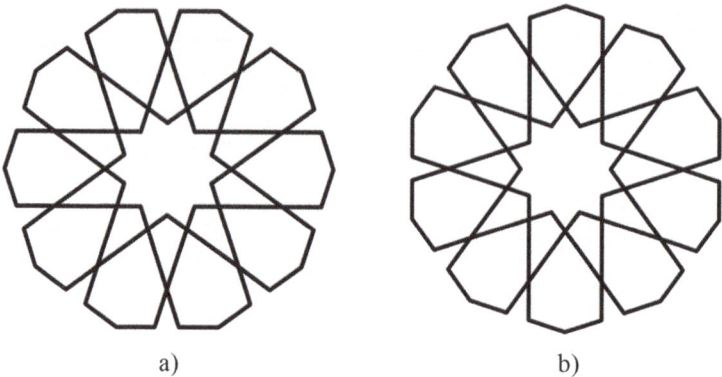

Fig. 3.38 Different visual properties depending on the dominant directions

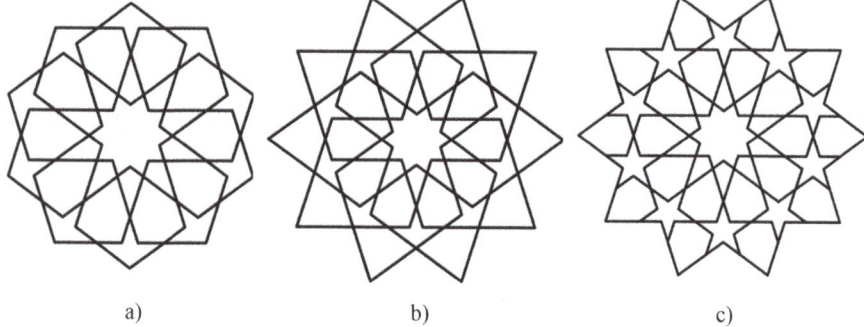

Fig. 3.39 Ornamentation of the exterior of a rosette with: (**a**) 1 layer; (**b**) 2 layers; (**c**) 3 layers

3.3.6 Ornamentation of a Rosette

It is customary to decorate the outside of a rosette by extending its sides, intersecting them, and connecting the found intersection points. Figure 3.39 shows different levels of ornamentation of the same rosette.

3.3.7 Coloring of a Rosette

Rosettes and its extensions have different aesthetic qualities depending on whether they are represented by lines or in the form of a mosaic coloring the different layers (Figs. 3.40 and 3.41).

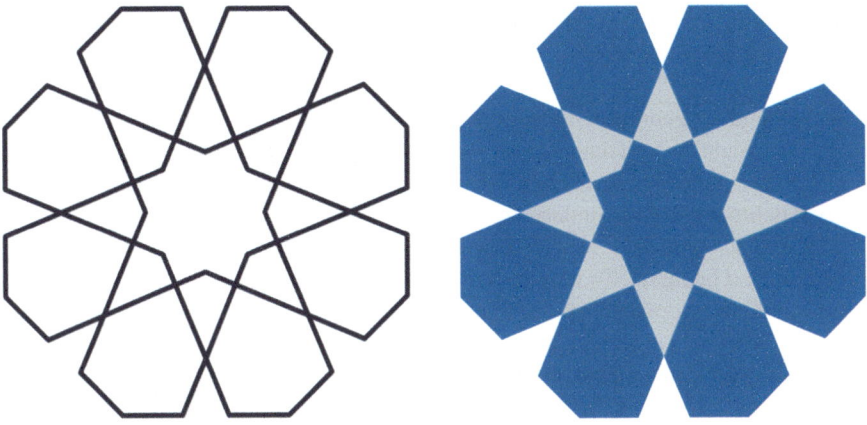

Fig. 3.40 Rosette represented by lines and in the form of colored mosaic

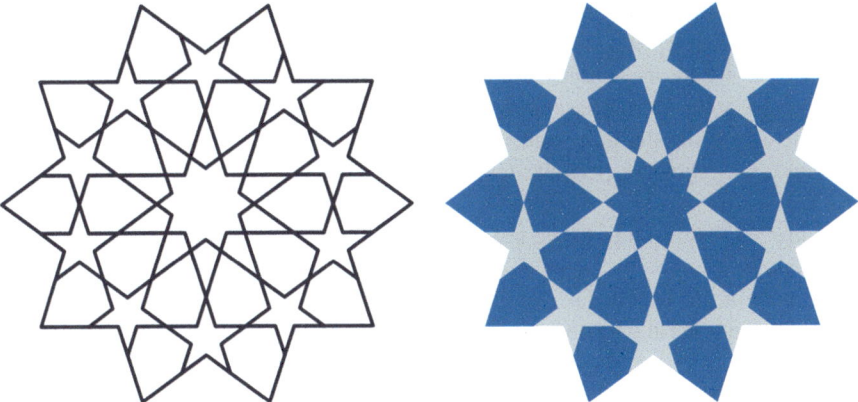

Fig. 3.41 Extended rosette represented by lines and in the form of colored mosaic

3.4 Tessellations with Stars and Rosettes

In this section, we present the radial extension method to generate periodic tessellations that have the repeat unit centrally decorated with one or more regular stars or rosettes, which are called central motifs. Then we describe how to draw them with interactive geometry software (IGS).

3.4.1 Determination of the Repeat Unit and, if Necessary, the Base Unit

To understand, in geometric terms, the design of a classic tessellation with stars or rosettes in the style of Islamic art, it is first necessary to determine precisely the single or several geometrically central motifs within the repeating unit and the polygon that delimits it, the boundary polygon, which is generally a rectangle, a square, or a regular hexagon and, less frequently, a parallelogram, a rhombus, or an elongated/flattened hexagon. To identify the boundary polygon is necessary to locate its vertices, which are points where at least three repeat units coincide. You have to identify four or six vertices, and it is not always easy: sometimes there is more than one possibility, in others there are missing or extra points, etc. Next, we will join the points located in cyclic order to form the boundary polygon, which will have four or six sides.

When the repeat unit contains more than one central motif, it is necessary to determine the base unit, the component of the repeat unit that contains only one central motif and which through axial symmetries, central symmetries, or rotations will populate the repeat unit. To identify the boundary polygon of the base unit is necessary to locate its vertices, similarly at the case of the repeat unit, which are points where at least three base units coincide. Normally, it will be a triangle, square, or isosceles/rectangular trapezoid.

Figure 3.42 shows a periodic tessellation with repeat unit containing a single rosette:

Figure 3.43 shows a periodic tessellation with repeat unit containing four regular stars:

Tessellations with regular stars and rosettes in the style of Islamic art can be constructed in several ways: with ruler and compass, with the grid method, with the technique called polygons in contact, or with the modular design system using a small set of basic geometric shapes, among others. Next, we will describe how to construct them using what we call the radial extension method.

3.4.2 Construction of Tessellations with Stars and Rosettes Using the Radial Extension Method

The construction of a tessellation using the radial extension method is done in two or three steps.

1. Draw the base motif of the repeat unit or base unit: a regular star or rosette.
2. Build the repeat unit or the base unit.

The part of the repeat unit or the base unit outside the base motif is called the interstitial region. The interstitial region is filled so that the repeat unit or the base unit parts bordering their adjacent ones in the final tessellation match correctly.

3.4 Tessellations with Stars and Rosettes

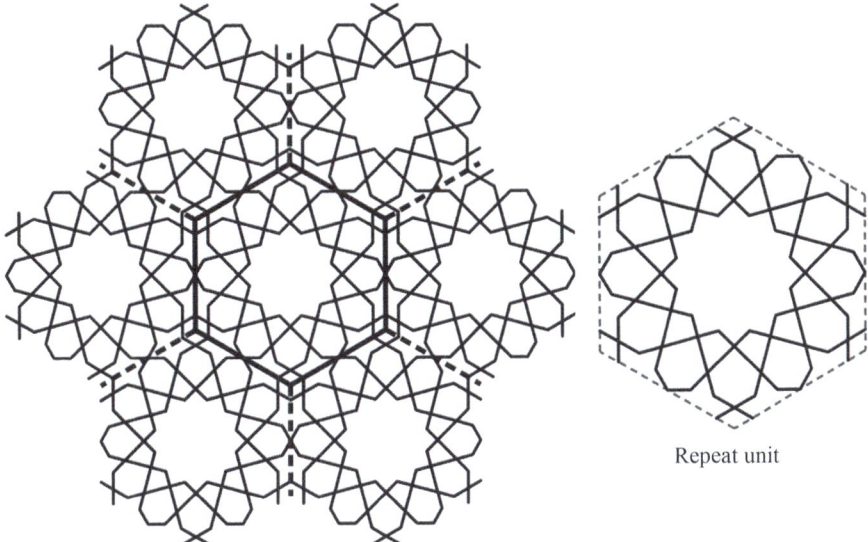

Fig. 3.42 Periodic tessellation with repeat unit containing a single rosette

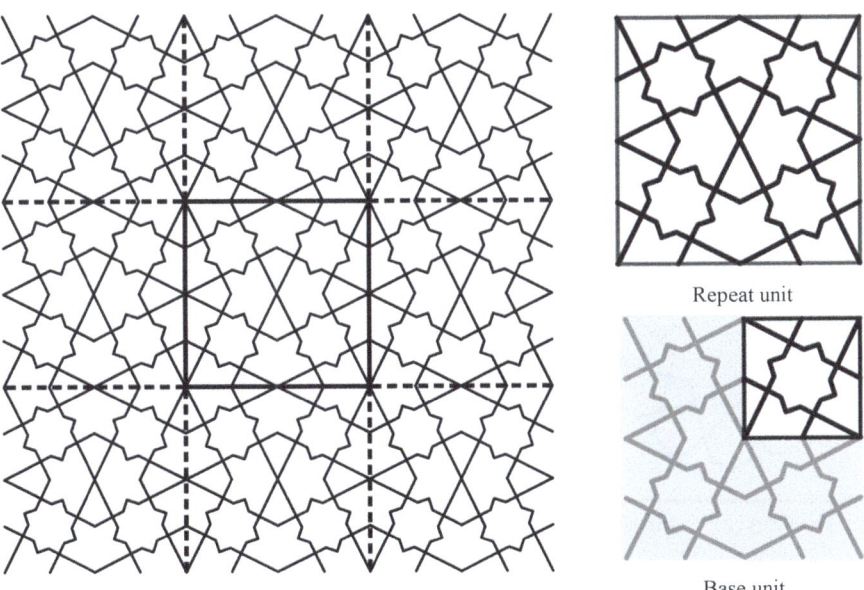

Fig. 3.43 Periodic tessellation with repeat unit containing four regular stars

In the simplest designs, the boundary polygon of the repeat unit or base unit is determined directly from the vertices and outermost sides of the base motif and in the most complex ones after the base motif expansion process in the interstitial region has started. The boundary polygon is a regular hexagon, square, rhombus, elongated hexagon, rectangle, or triangle. Once the boundary polygon is drawn, the extension of the base motif within the interstitial region is completed.

There is no single way to extend the base motif within the interstitial region. Generally, the outermost sides of the base motif are extended until they intersect with each other inside the pattern or with the boundary polygon of the repeat unit or the base unit. Then the interstitial region is filled by connecting the intersection points found. In the most complicated designs, it is sometimes necessary to draw: lines through two points, perpendiculars, parallels, tangents to a circle; rays extending the sides of stars and rosettes; circles given their center and passing through a point; intersections between rays, lines, and circles; midpoint of segments; perpendicular bisectors; rotations, axial, and central symmetries of points and lines, etc.

3. In case the repeat unit does not match the base unit, we construct the repeat unit from the base unit generally using axial and central symmetries.

The final periodic tessellation is created by multiple translations of the repeat unit.

Figures 3.44 and 3.45 are examples of the construction by the radial extension method of the repeat unit and the base unit of the tessellations of Figs. 3.42 and 3.43 whose central motifs are a rosette and a regular star, respectively. The use of color significantly enriches the design (see Figs. 3.46 and 3.47).

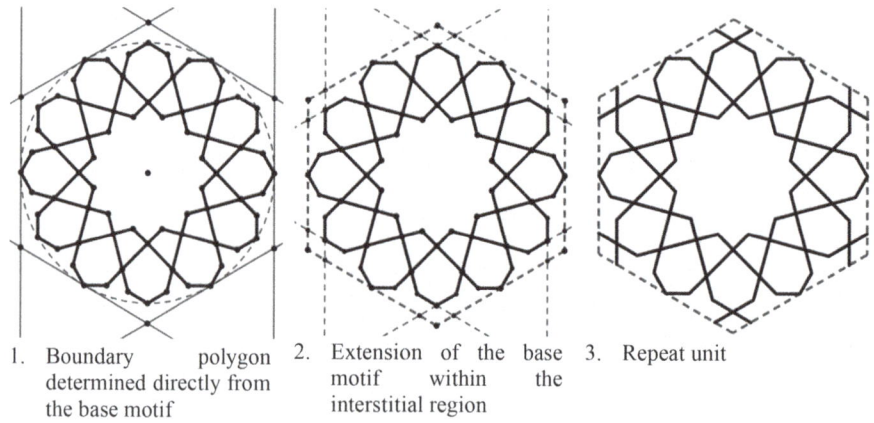

1. Boundary polygon determined directly from the base motif
2. Extension of the base motif within the interstitial region
3. Repeat unit

Fig. 3.44 Construction by the radial extension method of the repeat unit. 1. Boundary polygon determined directly from the base motif; 2. Extension of the base motif within the interstitial region; 3. Repeat unit

3.4 Tessellations with Stars and Rosettes

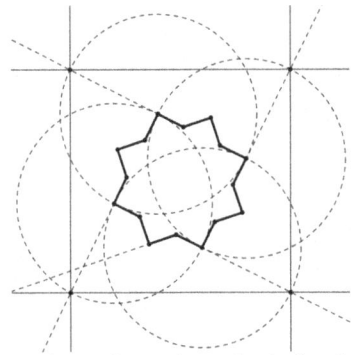
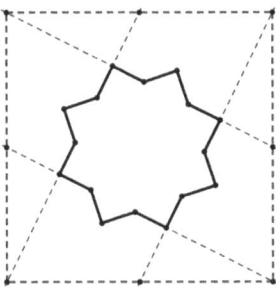

1. Boundary polygon determined after the base motif expansion process in the interstitial region has started
2. Extension of the base motif within the interstitial region

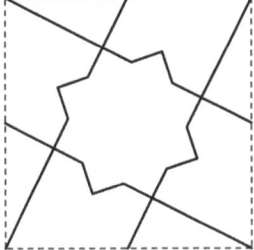

3. Base unit
4. Repeat unit

Fig. 3.45 Construction by the radial extension method of the base unit and the repeat unit. 1. Boundary polygon determined after the base motif expansion process in the interstitial region has started; 2. Extension of the base motif within the interstitial region; 3. Base unit; 4. Repeat unit

Fig. 3.46 Masjid Suleyman Pasha, Cairo, Egypt

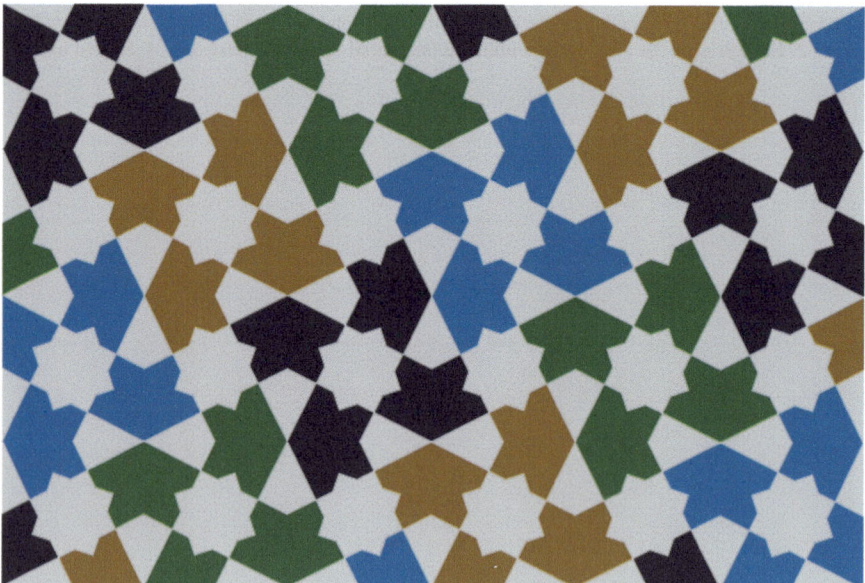

Fig. 3.47 Mosaic of the Alhambra Museum, Granada, Spain

Design of Tessellations with Stars and Rosettes

4.1 Introduction

This chapter presents 50 models of classic designs of periodic tessellations with stars and rosettes. For each model, a place where it can be found represented is indicated. Describes how to draw each of the models, step by step, with the radial extension method using Interactive Geometry Software (IGS).

Models 1–20 show how to draw the repeat unit when the base motif is a regular star and models 21–25 show how to draw the base unit when the base motif is a regular star.

Models 26–46 show how to draw the repeat unit when the base motif is a rosette and models 47–50 how to draw the base unit when the base motif is a rosette.

In each of the four groups, models are presented in increasing order according to the number of spikes on the regular star or the number of petals on the rosette used as the base motif. For each model, the level of difficulty of constructing the repeated unit or the base unit is indicated as: easy, medium, or difficult.

During the description of the generation of the repeat or the base unit of each model, only the operations that involve greater complexity will be given in detail. In the other cases, we will limit ourselves to drawing the different steps that allow us to complete the model.

4.1.1 Charles V Ceiling Room, Royal Alcazar, Seville, Spain

One can construct the tessellation of Fig. 4.1 using a repeat unit which is a square.

The design of Fig. 4.1 contains divergent 4-pointed, parallel 6-pointed, and divergent 8-pointed regular stars.

Fig. 4.1 Periodic tessellation structure of model 1

4.1 Introduction

Construction of the repeat unit

Level: Medium

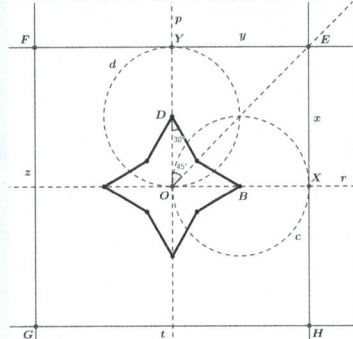
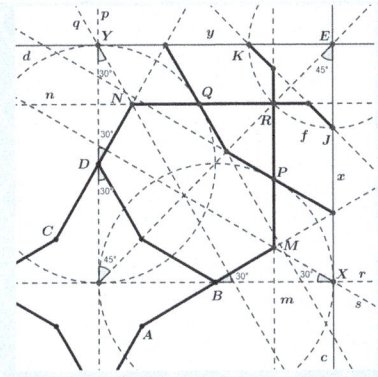

Drawing of the model 1. Step 1
1. Start from a (4,60°) = |4/(4/3)| regular star. Draw: the circle c of center B through O; the circle d of center D through O; the horizontal r through B; the vertical p through D; the point of intersection X of c and r; the point of intersection Y of d and p. Draw the square of vertices E, F, G, and H, boundary of the repeat unit, as shown.

Drawing of the model 1. Step 2
2. Draw: The line s rotation of r of center X and angle of 30°; the line q rotation of p of center Y and angle of 30°; the intersection M of the ray of origin A through B and s; the intersection N of the ray of origin C through D and q; the vertical m through M; the horizontal n through N; the intersection P of m and c; the intersection Q of n and d; the intersection R of m and n; the circle f of center E through R; the intersection J of line x and circle f; the intersection K of line y and circle f. finish the drawing as shown.

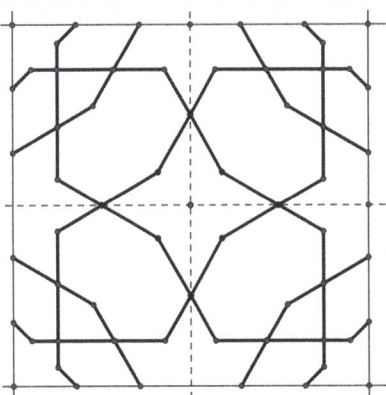
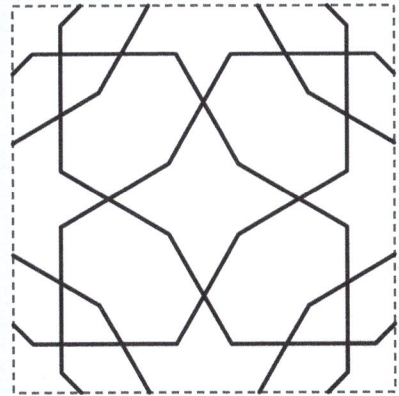

Drawing of the model 1. Step 3
3. Complete the drawing within the interstitial region of the repeat unit using axial symmetries.

Drawing of the model 1. Step 4
4. Repeat unit.

Fig. 4.2 Charles V Ceiling Room, Royal Alcazar, Seville, Spain

Obtain the final periodic tessellation by multiple translations of the repeat unit (Fig. 4.2).

The tessellation contains divergent $(4,60°) = |4/(4/3)|$, parallel $(6,60°) = |6/2|$, and convergent $(8,90°) = |8/2|$ regular stars.

4.1 Introduction

4.1.2 Nasrid Tripod Fountain, Alhambra, Granada, Spain

One can construct the tessellation of Fig. 4.3 using a repeat unit which is a regular hexagon.

The design of Fig. 4.3 contains divergent 6-pointed regular stars.

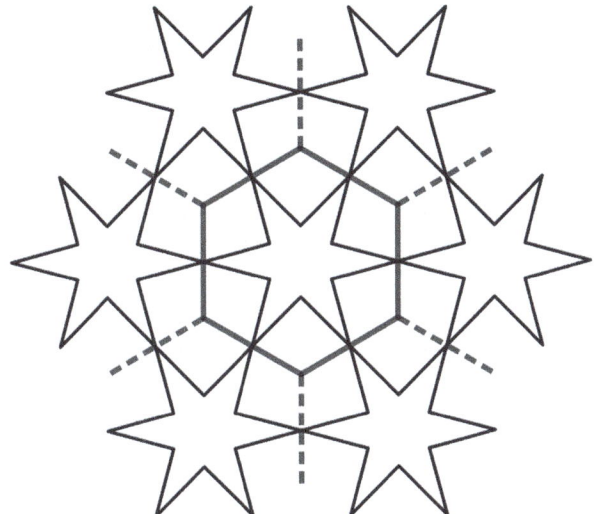

Fig. 4.3 Periodic tessellation structure of model 2

Construction of the repeat unit **Level: Easy**

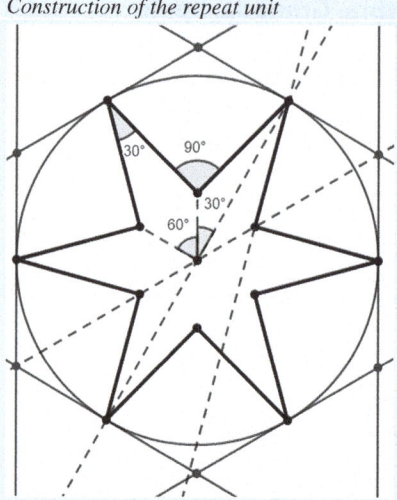 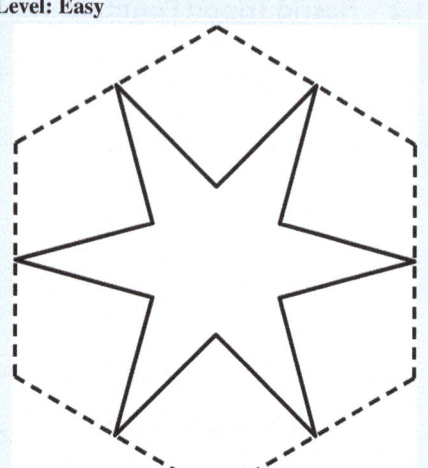

Drawing of the model 2. Step 1
1. Start from a divergent (6,30°) = |6/2.5| regular star. Draw the circle that passes through the spike vertices of the star. The boundary polygon is the regular hexagon bounded by the tangents to the circle at the spike vertices of the star.

Drawing of the model 2. Step 2
2. The repeat unit

4.1 Introduction

Fig. 4.4 Nasrid tripod fountain, Alhambra, Granada, Spain

Obtain the final periodic tessellation by multiple translations of the repeat unit (Fig. 4.4).

4.1.3 Tash Hauli Palace Complex, Khiva, Uzbekistan

One can construct the tessellation of Fig. 4.5 using a repeat unit which is a regular hexagon.

The design of Fig. 4.5 contains convergent 6-pointed regular stars.

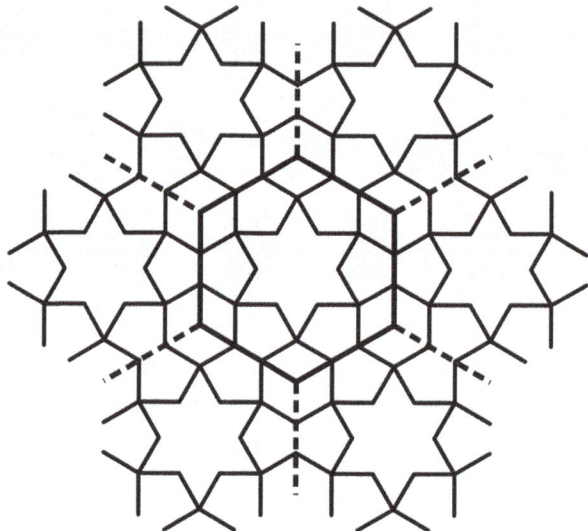

Fig. 4.5 Periodic tessellation structure of model 3

4.1 Introduction

Construction of the repeat unit **Level: Medium**

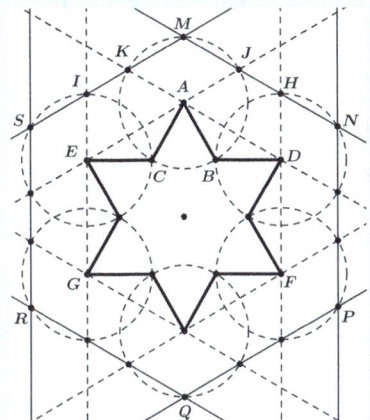
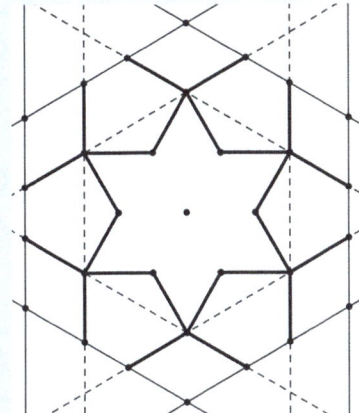

Drawing of the model 3. Step 1
1. Start from a (6,60°) = |6/2| regular star. Draw: The point J intersection of the circle with center point A passing through the point B and the line passing through the points A and E; the point of intersection K of the circle with center A passing through the point C and the line passing through the points A and D; the point H intersection of the circle with center D passing through B and the line passing through the points F and D; the point I intersection of the circle with center the point E passing through C and the line passing through the points G and E; the point M intersection of the lines passing through the points H and J and through I and K. Draw in the same way the points N, P, Q, R, and S that together with M determine the boundary of the hexagonal repeated unit.

Drawing of the model 3. Step 2
2. Complete the drawing within the interstitial region of the repeat unit.

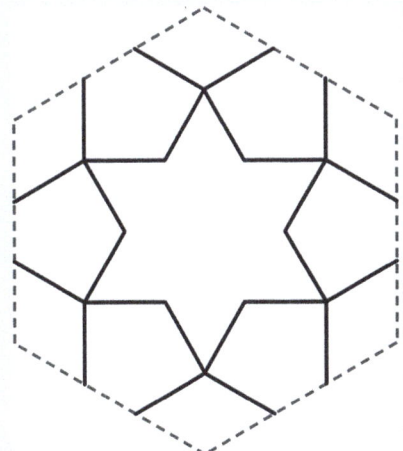

Drawing of the model 3. Step 3
3. The repeat unit.

Fig. 4.6 Tash Hauli Palace Complex, Khiva, Uzbekistan

Obtain the final periodic tessellation by multiple translations of the repeat unit (Fig. 4.6).

4.1 Introduction

4.1.4 Court of the Myrtles, Alhambra, Granada, Spain

One can construct the tessellation of Fig. 4.7 using a repeat unit which is a square. The design of Fig. 4.7 contains convergent 8-pointed regular stars.

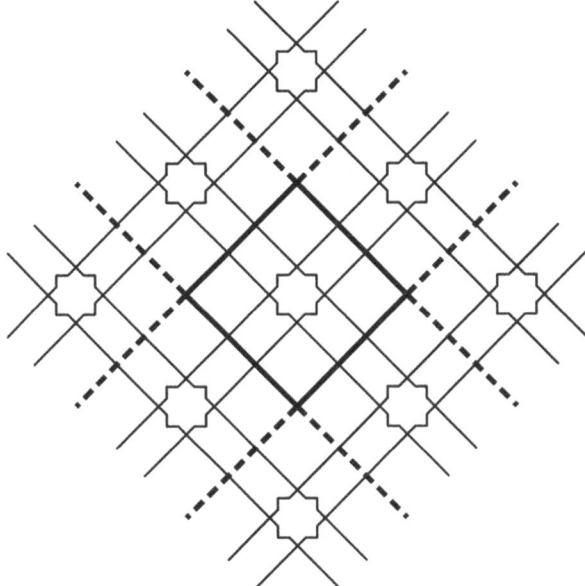

Fig. 4.7 Periodic tessellation structure of model 4

4 Design of Tessellations with Stars and Rosettes

Construction of the repeat unit **Level: Easy**

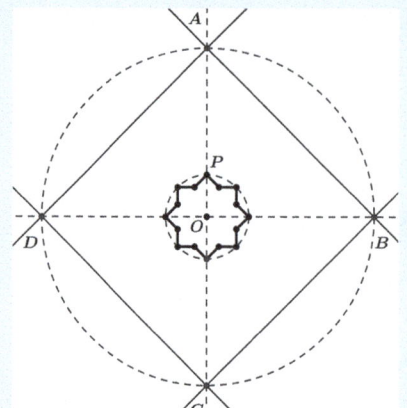
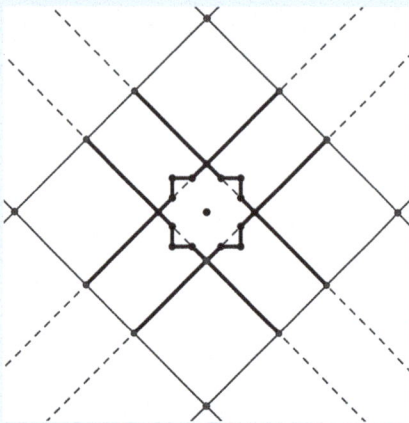

Drawing of the model 4. Step 1
1. Start from a (8,90°) = |8/2| regular star. Draw the large circle so that its radius is 4 times the radius of the small circle. Draw the vertices A, B, C, and D of the boundary of the repeat unit as shown.

Drawing of the model 4. Step 2
2. Expand some of the sides of the star and draw the points where they intersect the boundary polygon. Draw a part of the model as shown. Complete the drawing within the interstitial region to obtain the repeat unit.

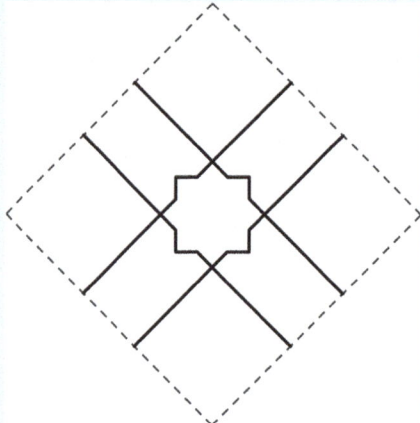

Drawing of the model 4. Step 3
3. Repeat unit.

4.1 Introduction

Fig. 4.8 Court of the Myrtles, Alhambra, Granada, Spain

Obtain the final periodic tessellation by multiple translations of the repeat unit (Fig. 4.8).

4.1.5 Nasrid Palace, Alhambra Museum, Granada, Spain

One can construct the tessellation of Fig. 4.9 using a repeat unit which is a square. The design of Fig. 4.9 contains ideal convergent 8-pointed regular stars.

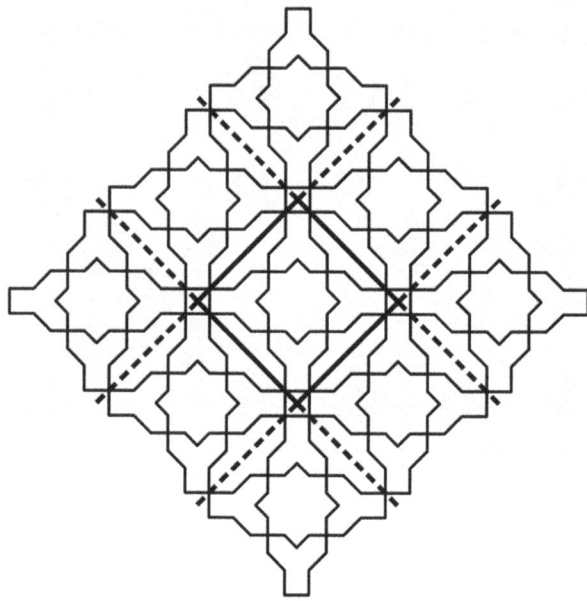

Fig. 4.9 Periodic tessellation structure of model 5

4.1 Introduction

Construction of the repeat unit   **Level: Medium**

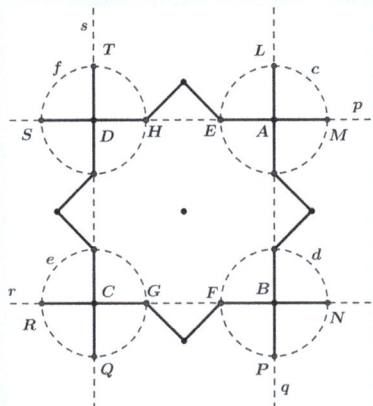 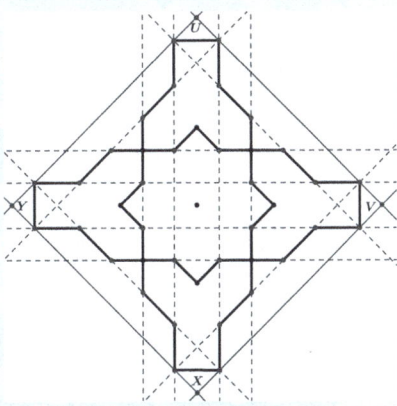

Drawing of the model 5. Step 1
1. Start from a (8,90°) = |8/2| regular star. Draw the circles c, d, e, and f of center A, B, C, and D passing through the points E, F, G, and H, respectively. Draw the lines p, q, r, and s through D and A, A and B, B and C, C and D, respectively. Draw: The point L intersection of the circle c and the line q; the point M intersection of the circle c and the line p; the point N intersection of the circle d and the line r; the point P intersection of the circle d and the line q; the point Q intersection of the circle e and the line s; the point R intersection of the circle e and the line r; the point S intersection of the circle f and the line p; the point T intersection of the circle f and the line s. Draw a part of the model as shown.

Drawing of the model 5. Step 2
2. Draw the lines and points of intersection as shown. Draw the points U, V, X, and Y that are the vertices of the square repeat unit. Draw the missing part of the model.

Drawing of the model 5. Step 3
3. Repeat unit.

Fig. 4.10 Nasrid Palace, Alhambra Museum, Granada, Spain

Obtain the final periodic tessellation by multiple translations of the repeat unit (Fig. 4.10).

4.1 Introduction

4.1.6 Royal Alcazar, Seville, Spain

One can construct the tessellation of Fig. 4.11 using a repeat unit which is a square. The design of Fig. 4.11 contains parallel and convergent 8-pointed regular stars.

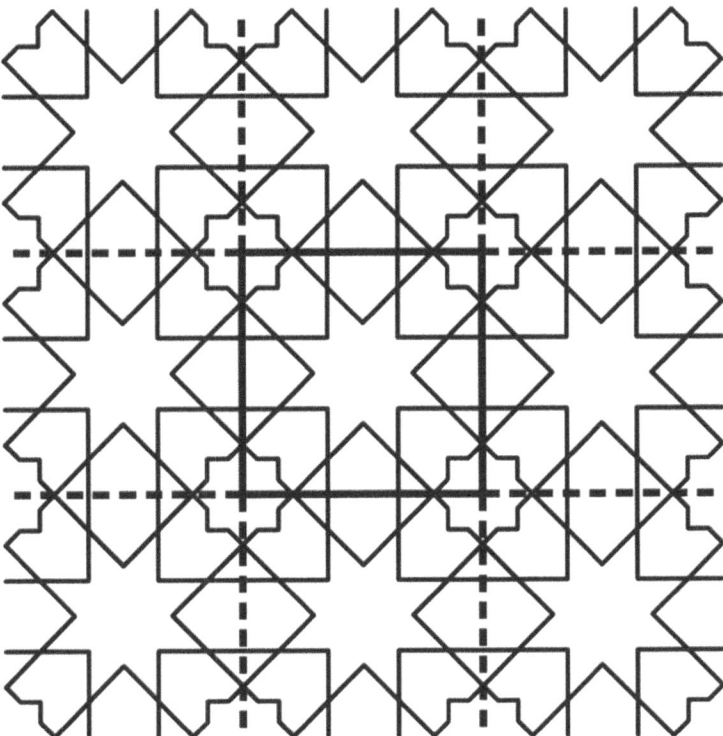

Fig. 4.11 Periodic tessellation structure of model 6

Construction of the repeat unit   **Level: Easy**

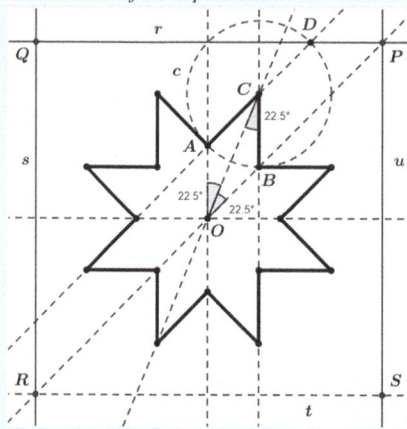
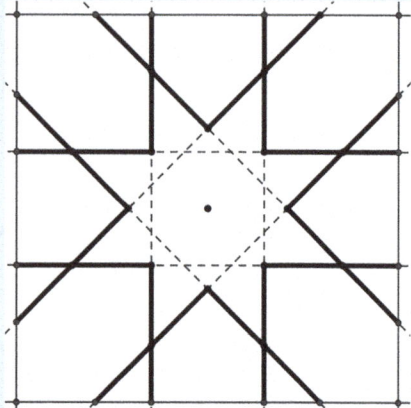

Drawing of the model 6. Step 1
1. Start from a parallel (8,45°) = |8/3| regular star. Draw: The circle c of center C passing through points A and B; the point D intersection of c with the line through A and C; the horizontal r through D; the lines s, t, and u obtained from r by successive rotations of center O and 90°; the vertices P, Q, R, and S of the boundary of the repeat unit as shown.

Drawing of the model 6. Step 2
2. Draw: The expanded sides of the star; the points where the expanded sides intersect the boundary of the repeat unit. Complete part of the design within the interstitial region as shown.

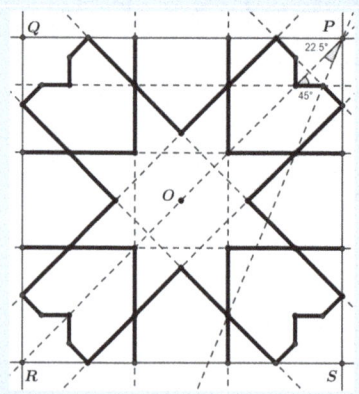

Drawing of the model 6. Step 3
3. Draw the lower left part of a (8.90°) = |8/2| regular star centered at point P as shown. Complete the drawing of the repeat unit with successive rotations of center O and 90° of part of the star of center P.

Drawing of the model 6. Step 4
4. Repeat unit.

4.1 Introduction

Fig. 4.12 Royal Alcazar, Seville, Spain

Obtain the final periodic tessellation by multiple translations of the repeat unit (Fig. 4.12).

The tessellation contains parallel $(8,45°) = |8/3|$ and convergent $(8,90°) = |8/2|$ regular stars.

4.1.7 Bibi-Khanum Mosque, Samarkand, Uzbekistan

One can construct the tessellation of Fig. 4.13 using a repeat unit which is a square.

The design of Fig. 4.13 contains convergent 8-pointed regular stars and divergent standard 4-pointed rosettes.

Fig. 4.13 Periodic tessellation structure of model 7

4.1 Introduction

Construction of the repeat unit   **Level: Medium**

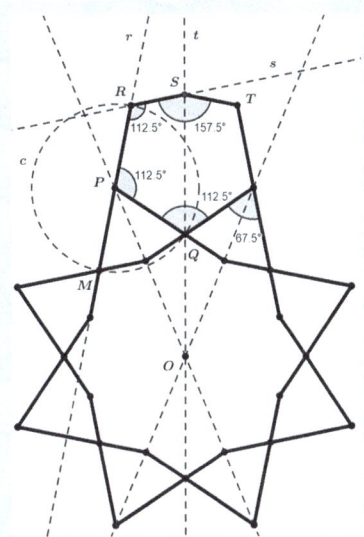

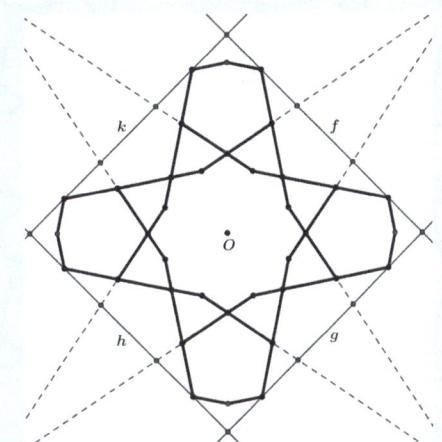

Drawing of the model 7. Step 1
1. Start from a convergent regular star (8,67.5°)2 = |8/2.5|2. Draw: The circle c of center P passing through the point Q; the point R intersection of c with the line r passing through the points M and P; the rotation s of r of center R and angle of 112.5° clockwise; the point S intersection of the line s and the line t passing through O and Q; the symmetric T of R with respect to t. Draw the hexagon on top of the star.

Drawing of the model 7. Step 2
2. Draw three more hexagons on the left, right, and bottom parts of the star using successive rotations of center O and 90° of the hexagon drawn on top. Draw the boundary polygon delimited by the lines f, g, h, and k. extend the sides of the star and cut them with the boundary polygon.

Drawing of the model 7. Step 3
3. Complete the drawing within the interstitial region of the repeat unit.

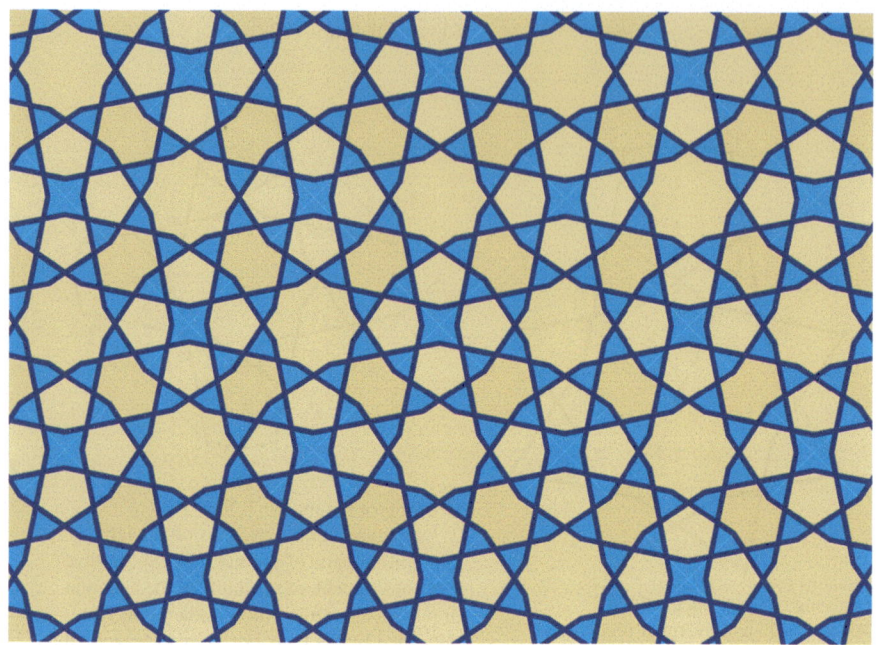

Fig. 4.14 Bibi-Khanum Mosque, Samarkand, Uzbekistan

Obtain the final periodic tessellation by multiple translations of the repeat unit (Fig. 4.14).

The tessellation contains convergent (8,67.5°)2 = |8/2.5|2 regular stars and divergent standard 4-pointed rosettes with center a (4,67.5°) = |4/1.25| regular star.

4.1 Introduction

4.1.8 Shah-i Zinda, Samarkand, Uzbekistan

One can construct the tessellation of Fig. 4.15 using a repeat unit which is a rectangle.

The design of Fig. 4.15 contains 8-pointed convergent stars.

Fig. 4.15 Periodic tessellation structure of model 8

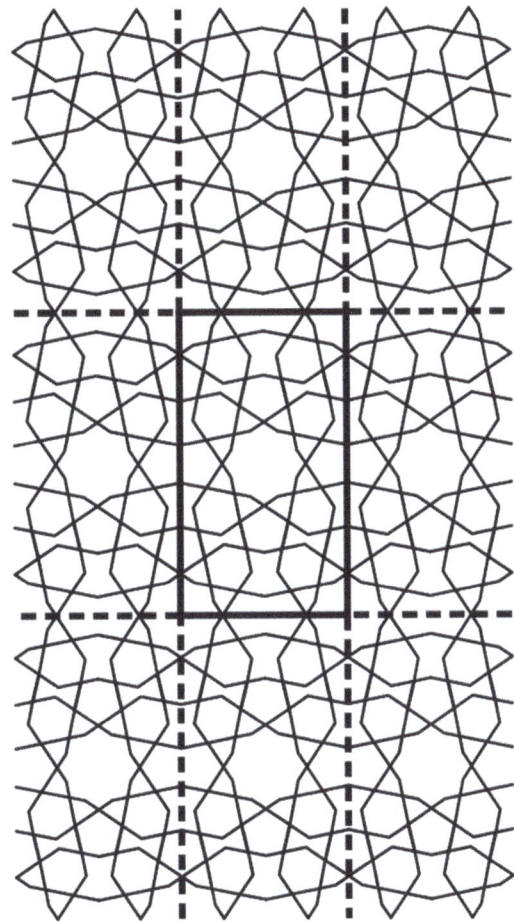

Construction of the repeat unit **Level: Medium**

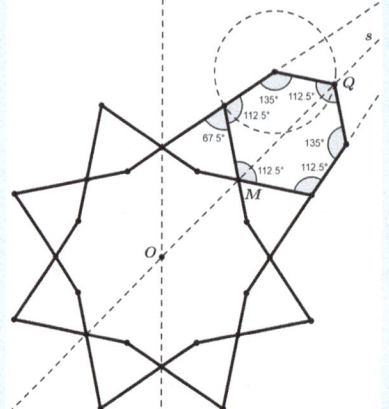
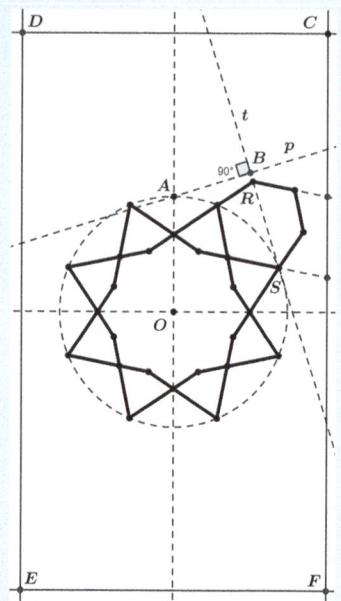

Drawing of the model 8. Step 1
1. Start from a convergent regular star $(8, 67.5°)2 = |8/2.5|2$. Draw a standard petal by selecting the point Q on the line s passing through the points O and M so that the interior angle at point Q of s measures 112.5°.

Drawing of the model 8. Step 2
2. Draw: The point A intersection of the circle passing through the vertices of the star and the vertical line passing through the point O; the line t passing through the points R and S; the perpendicular p to t passing through the point A; the point B intersection of p and t; the point C symmetric of the point O with respect to the point B. Draw the boundary polygon of the repeat unit of vertices C, D, E, and F as shown.

4.1 Introduction

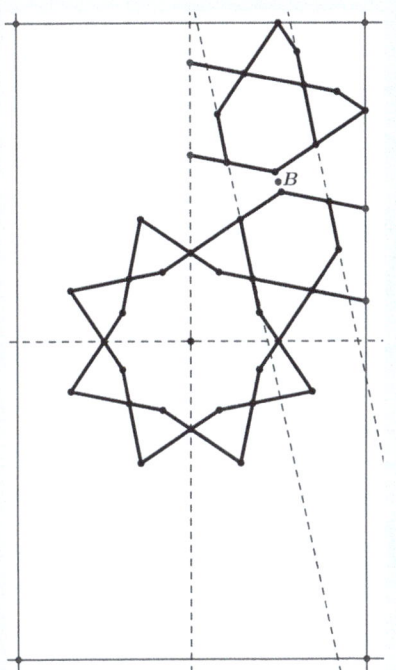

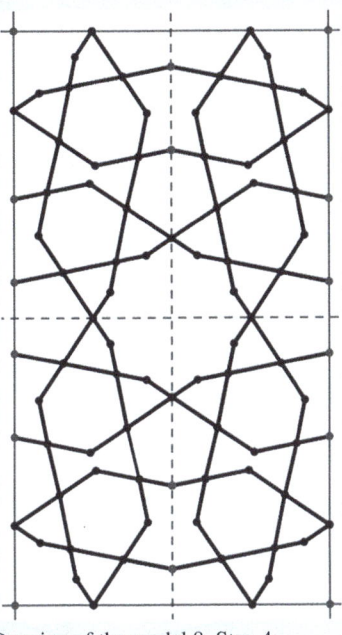

Drawing of the model 8. Step 3
3. The top of the upper right quadrant of the repeat unit is symmetrical to the bottom with respect to point B.

Drawing of the model 8. Step 4
4. Complete the drawing within the interstitial region of the repeat unit using axial symmetries.

Drawing of the model 8. Step 5

5. Repeat unit.

4.1 Introduction

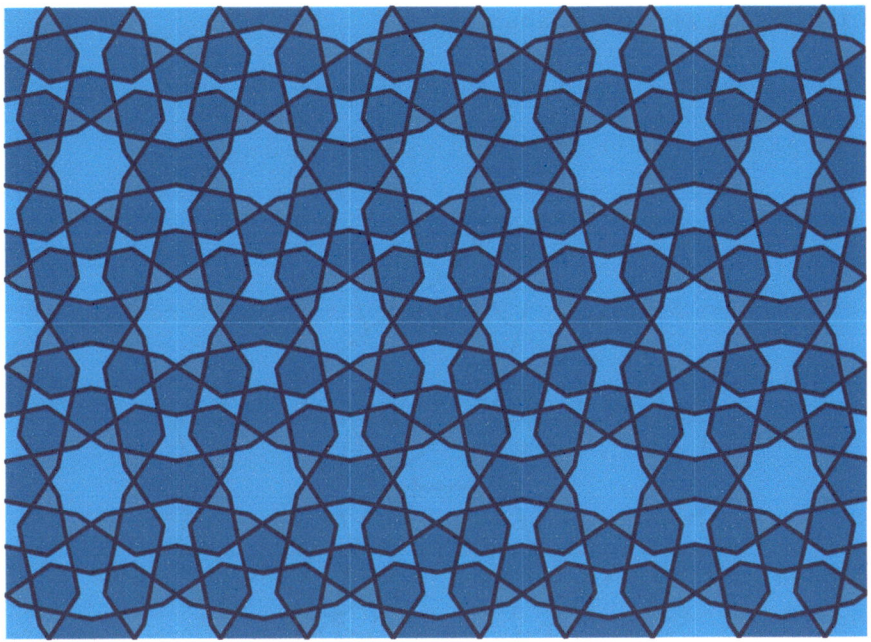

Fig. 4.16 Shah-i Zinda, Samarkand, Uzbekistan

Obtain the final periodic tessellation by multiple translations of the repeat unit (Fig. 4.16).

4.1.9 Friday Mosque, Isfahan, Iran

One can construct the tessellation of Fig. 4.17 using a repeat unit which is an elongated hexagon.

The design of Fig. 4.17 contains convergent 10-pointed regular stars.

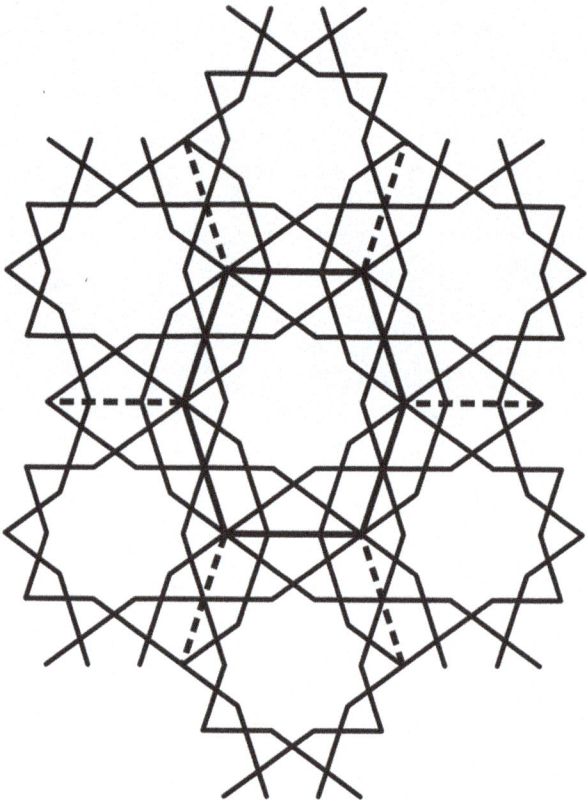

Fig. 4.17 Periodic tessellation structure of model 9

4.1 Introduction

Construction of the repeat unit **Level: Easy**

Drawing of the model 9. Step 1
1. Start from a convergent regular star $(10, 72°)2 = |10/3|2$.

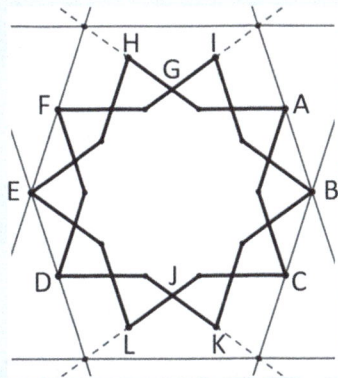

Drawing of the model 9. Step 2
2. The boundary f of the repeat unit is the elongated hexagon of vertices: The point of intersection of the line through A and B with the ray of origin G through I; point B; the point of intersection of the line through B and C with the ray of origin J through K; the point of intersection of the line through D and E with the ray of origin J through L; point E; the point of intersection of the line through E and F with the ray of origin G through H.

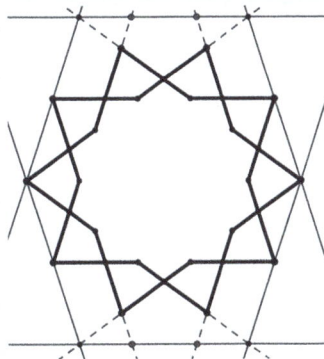

Drawing of the model 9. Step 3
3. Extend the sides of the vertices of the star that do not belong to the boundary f and find the points where they intersect it.

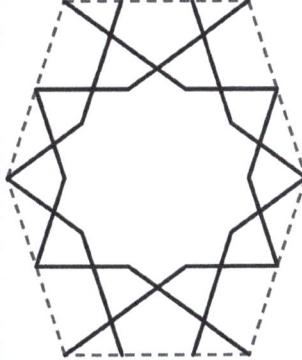

Drawing of the model 9. Step 4
4. Complete the drawing within the interstitial region of the repeat unit.

Fig. 4.18 Friday Mosque, Isfahan, Iran

Obtain the final periodic tessellation by multiple translations of the repeat unit (Fig. 4.18).

4.1 Introduction

4.1.10 Akbar's Mausoleum, Sikandra, India

One can construct the tessellation of Fig. 4.19 using a repeat unit which is a rectangle.

The design of Fig. 4.19 contains convergent 10-pointed and divergent 5-pointed regular stars.

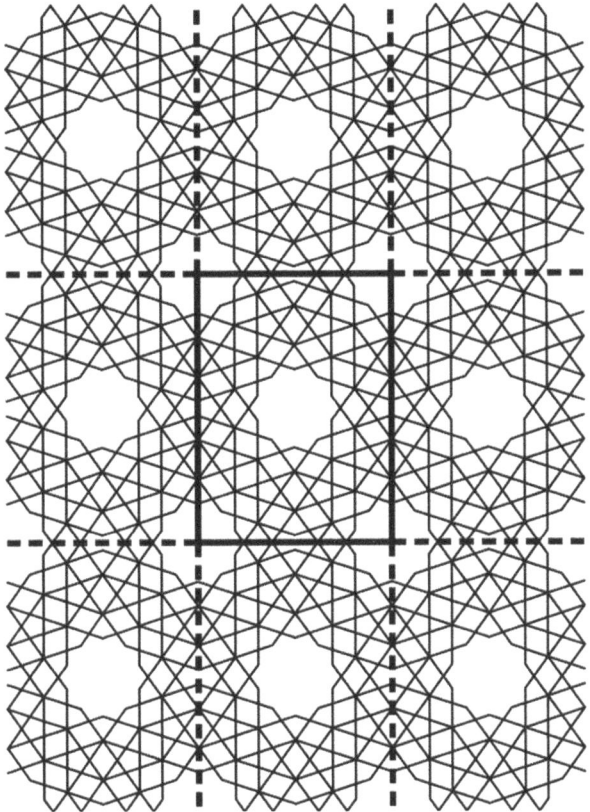

Fig. 4.19 Periodic tessellation structure of model 10

Construction of the repeat unit **Level: Medium**

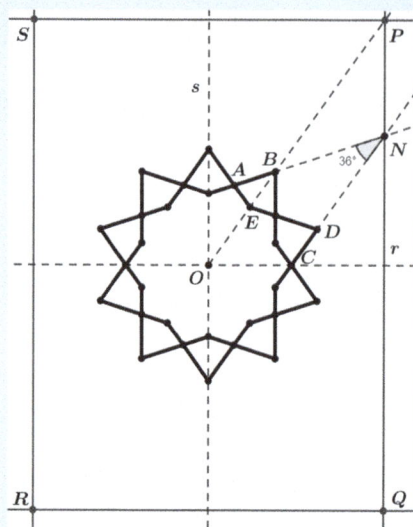

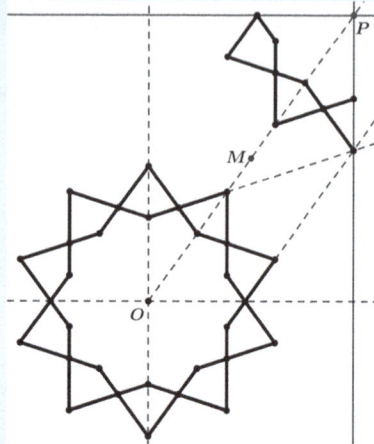

Drawing of the model 10. Step 1
1. Start from a convergent $(10,72°)2 = |10/3|2$ regular star. Draw: The point N intersection of the ray of origin A through B and the ray of origin C through D; the point P intersection of the vertical through N and the ray of origin O through E; the points Q, R, and S using axial symmetries with respect to the lines r and s. the boundary polygon of the repeat unit is the rectangle of vertices P, Q, R, and S.

Drawing of the model 10. Step 2
2. Draw: The midpoint M of O and P; the symmetrical with respect to M of the lower part of the upper right quadrant as shown.

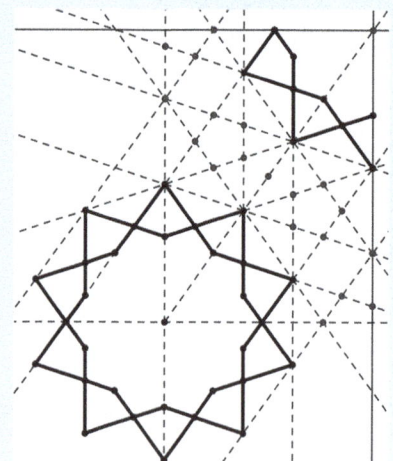

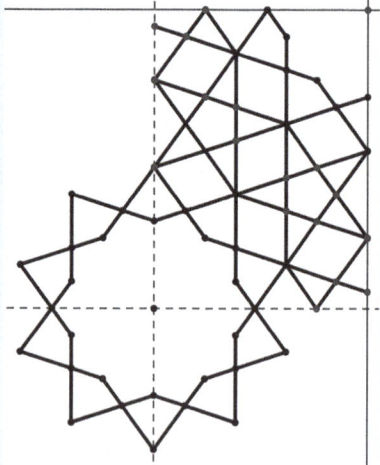

Drawing of the model 10. Step 3
3. Draw rays, lines, and intersection points within the upper right part of the interstitial region of the repeat unit as shown.

Drawing of the model 10. Step 4
4. Complete the drawing within the upper right part of the interstitial region of the repeat unit as shown.

4.1 Introduction

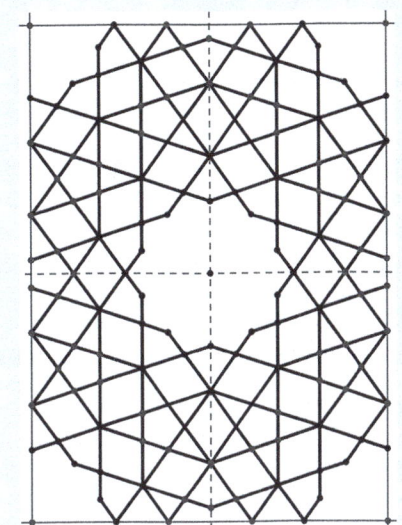

Drawing of the model 10. Step 5
5. Complete the drawing within the interstitial region of the repeat unit using axial symmetries.

Drawing of the model 10. Step 6
6. The repeat unit.

Fig. 4.20 Akbar's Mausoleum, Sikandra, India

Obtain the final periodic tessellation by multiple translations of the repeat unit (Fig. 4.20).

The tessellation contains convergent $(10,72°)2 = |10/3|2$ regular stars and divergent $(5,36°) = |5/2|$ regular stars with a central regular pentagon.

4.1 Introduction

4.1.11 Al-Maridani Mosque, Cairo, Egypt

One can construct the tessellation of Fig. 4.21 using a repeat unit which is a rectangle. The design of Fig. 4.21 contains convergent 10-pointed regular stars.

Fig. 4.21 Periodic tessellation structure of model 11

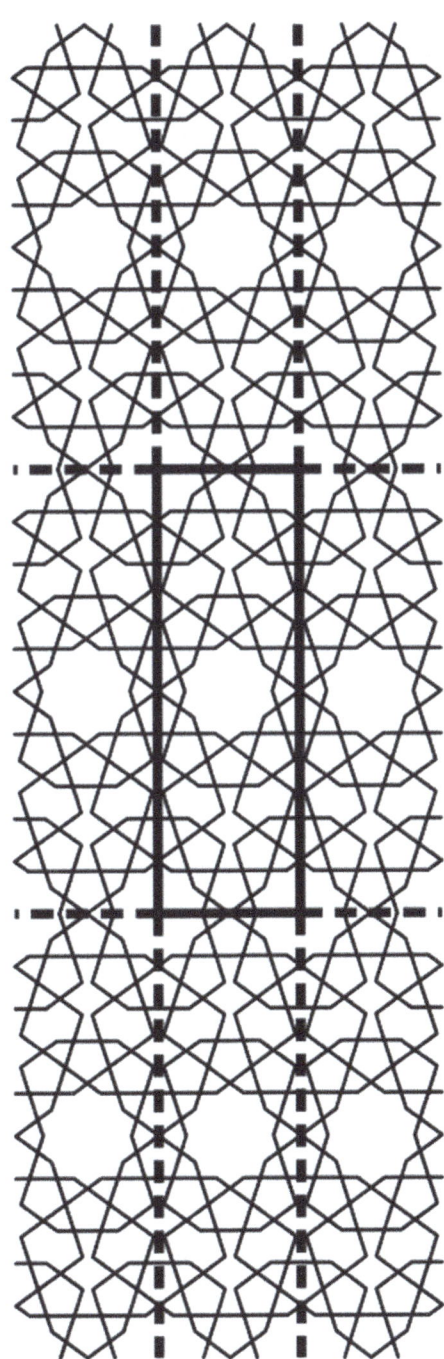

Construction of the repeat unit   **Level: Difficult**

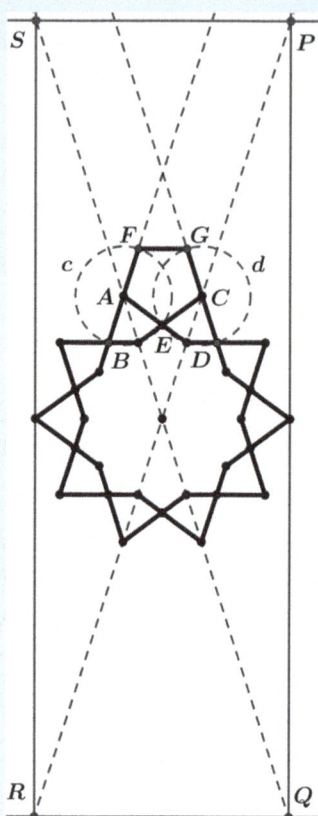

Drawing of the model 11. Step 1
1. Start from a convergent $(10,72°)2 = |10/3|2$ regular star. Draw the boundary polygon of the repeat unit of vertices P, Q, R, and S as shown. Draw: The circle c of center A through E; the circle d of center C through E; the point of intersection F of the circle c and the ray of origin B through A; the point of intersection G of the circle d and the ray of origin D through C. Draw the regular pentagon AFGCE.

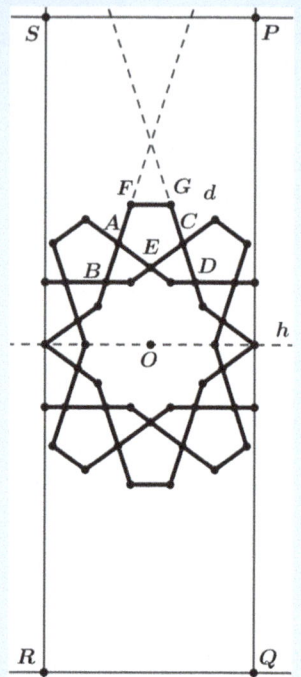

Drawing of the model 11. Step 2
2. Rotate the pentagon AFGCE with rotations of center O and angle of 36° clockwise and counterclockwise. Draw the lower pentagons with a symmetry of the upper pentagons of axis the horizontal line h passing through O.

4.1 Introduction

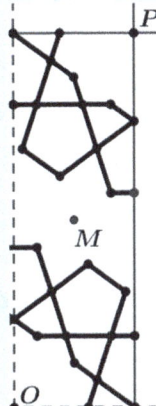

Drawing of the model 11. Step 3
3. Draw the midpoint M of O and P. Draw the upper part symmetrical with respect to M of the lower part as shown.

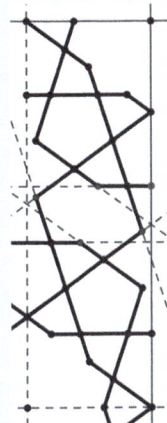

Drawing of the model 11. Step 4
4. Complete the drawing within the upper right part of the interstitial region of the repeat unit as shown.

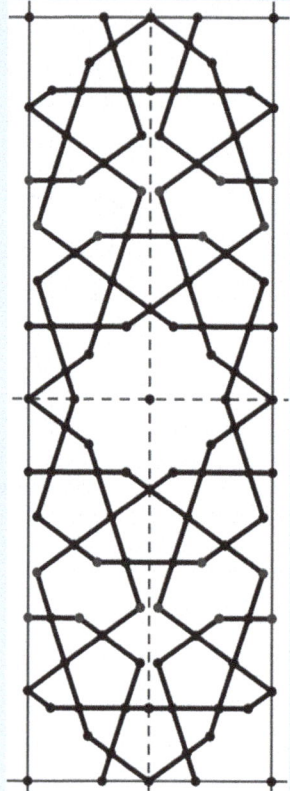

Drawing of the model 11. Step 5
5. Complete the drawing within the interstitial region of the repeat unit using axial symmetries.

Drawing of the model 11. Step 6
6. Repeat unit.

4.1 Introduction

Fig. 4.22 Al-Maridani Mosque, Cairo, Egypt

Obtain the final periodic tessellation by multiple translations of the repeat unit (Fig. 4.22).

4.1.12 Bibi-Khanum Mosque, Samarkand, Uzbekistan

One can construct the tessellation of Fig. 4.23 using a repeat unit which is a rectangle.

The design of Fig. 4.23 contains convergent 10-pointed regular stars.

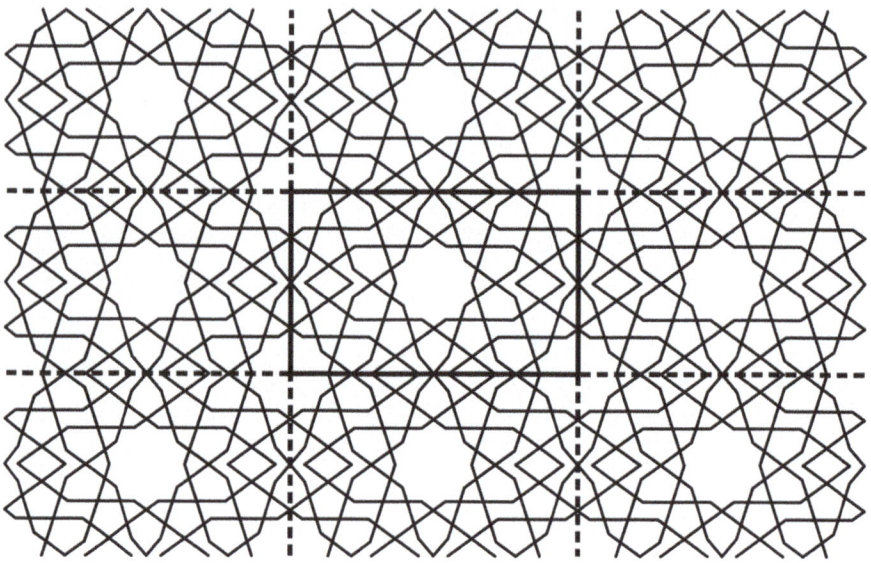

Fig. 4.23 Periodic tessellation structure of model 12

4.1 Introduction

Construction of the repeat unit **Level: Difficult**

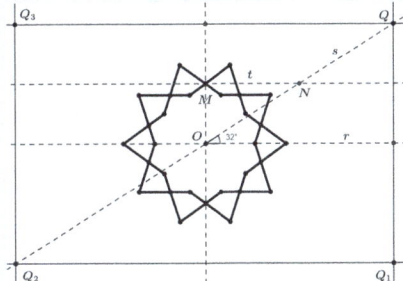

Drawing of the model 12. Step 1
1. Start from a convergent (10,72°)2 = |10/3|2 regular star. Draw: The line s rotation of the line r of center O and angle 32°; the point N of intersection of the lines s and the horizontal through the point M; the point Q symmetric of the point O with respect to the point N.

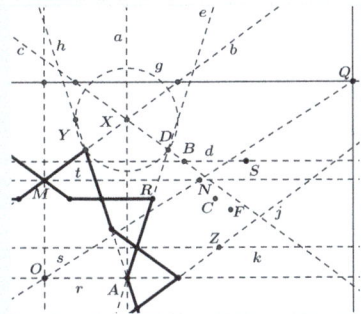

Drawing of the model 12. Step 2
2. Draw: The point S symmetric of the point R with respect to the point N; the vertical a through the point A; the line c symmetric of the line b with respect to the line a; the horizontal d through the point S; the point of intersection B of the lines c and d; the point C symmetric of the point B with respect to the point N; the point D intersection of the lines c and e; the point F symmetric of the point D with respect to the point N; the point X intersection of the lines b and c; the circle g of center the point X through the point D; the point Y intersection of the line h and the circle g.

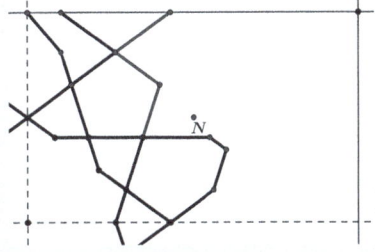

Drawing of the model 12. Step 3
3. Draw the left part of the model located in the upper right quadrant of the repeat unit.

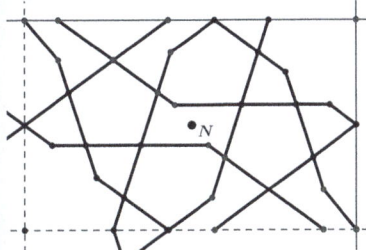

Drawing of the model 12. Step 4
4. The right part of the upper right quadrant of the repeating unit is symmetrical to the left part with respect to the point N.

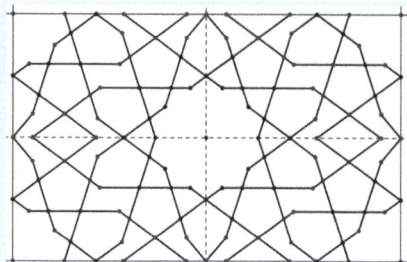 

Drawing of the model 12. Step 5
5. Complete the design using axial symmetries.

Drawing of the model 12. Step 6
6. Repeat unit.

4.1 Introduction

Fig. 4.24 Bibi-Khanum Mosque, Samarkand, Uzbekistan

Obtain the final periodic tessellation by multiple translations of the repeat unit (Fig. 4.24).

4.1.13 Friday Mosque, Yazd, Iran

One can construct the tessellation of Fig. 4.25 using a repeat unit which is a rectangle.

The design of Fig. 4.25 contains two different types of regular convergent 10-pointed stars.

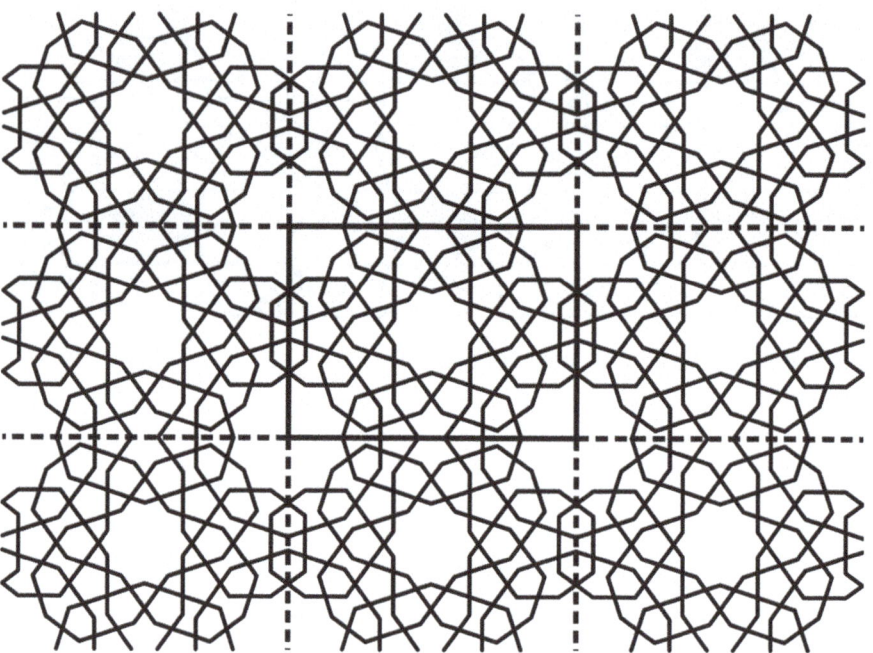

Fig. 4.25 Periodic tessellation structure of model 13

4.1 Introduction

Construction of the repeat unit   **Level: Difficult**

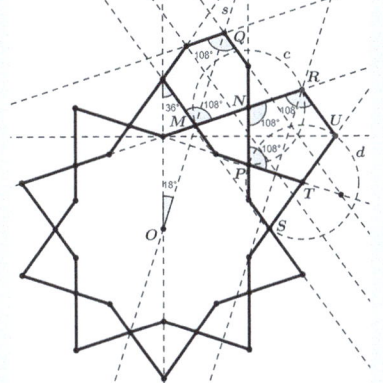

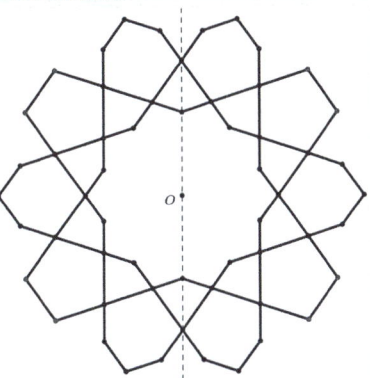

Drawing of the model 13. Step 1
1. Start from a convergent $(10,72°)2 = |10/3|2$ regular star. Draw a standard petal by selecting the point Q on the line s passing through the points O and M so that the interior angle at point Q measures 108°. Draw: The circle c of center N through M; the circle d of center T through S; the point of intersection R of the circle c and the ray of origin M through N; the point of intersection U of the circle d and the ray of origin S through T. Draw the regular pentagon of vertices P, N, R, U, and T.

Drawing of the model 13. Step 2
2. Draw the right side of the design with two successive rotations of the petal of center O and angle 72° clockwise and one rotation of the pentagon of center O and angle 72° clockwise. Complete the left part of the design using a symmetry of axis the vertical line through O.

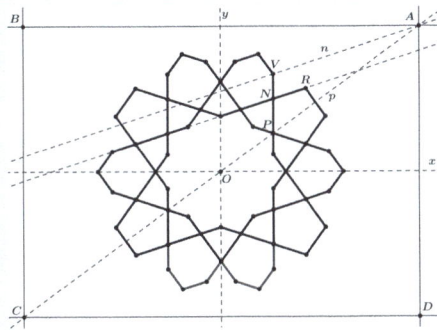

Drawing of the model 13. Step 3
3. Draw: The line p passing through the points O and P; the line n passing through the point V parallel to the line passing through the points N and R; the point A intersection of the lines p and n; the point B symmetrical of A with respect to the line x; the points C and D symmetrical of points B and A with respect to the line y, respectively. The rectangle of vertices A, B, C, and D is the boundary polygon of the repeat unit.

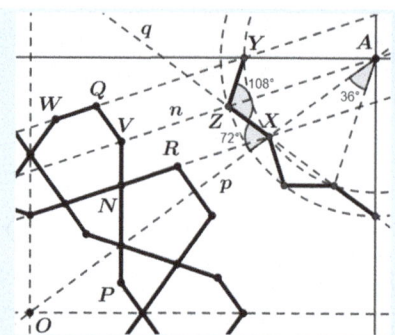
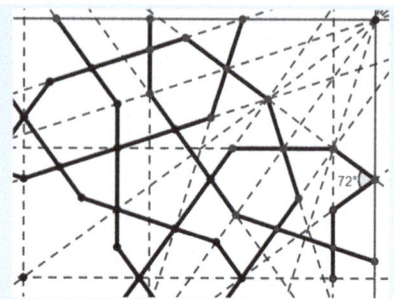

Drawing of the model 13. Step 4

Drawing of the model 13. Step 5

4. Draw: The point X intersection of the line p and the line passing through N and R; the point Y intersection of the line passing through the points W and Q and the horizontal through A; the line q rotated of p of center X and angle 72° clockwise; the point Z intersection of n and q; complete the part of the regular star $|10/2| = (10,108°)$ located at the top right of the interstitial region with rotations of center A and angle of 36° as shown.

5. Complete the drawing within the upper right part of the interstitial region of the repeat unit as shown.

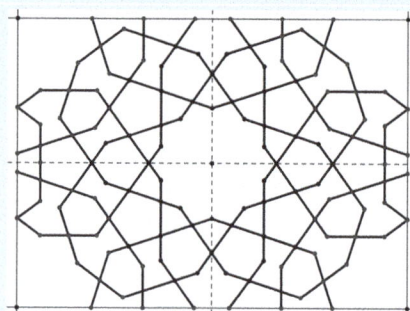

Drawing of the model 13. Step 6

Drawing of the model 13. Step 7

6. Complete the drawing within the interstitial region of the repeat unit using axial symmetries.

7. Repeat unit.

4.1 Introduction

Fig. 4.26 Friday Mosque, Yazd, Iran

Obtain the final periodic tessellation by multiple translations of the repeat unit (Fig. 4.26).

The tessellation contains convergent $(10,72°)2 = |10/3|2$ and $(10,108°) = |10/2|$ regular stars.

4.1.14 Abdullah Khan Madrasa, Bukhara, Uzbekistan

One can construct the tessellation of Fig. 4.27 using a repeat unit which is a rhombus.

The design of Fig. 4.27 contains convergent 10-pointed and convergent 5-pointed regular stars.

Fig. 4.27 Periodic tessellation structure of model 14

4.1 Introduction

Construction of the repeat unit   **Level: Medium**

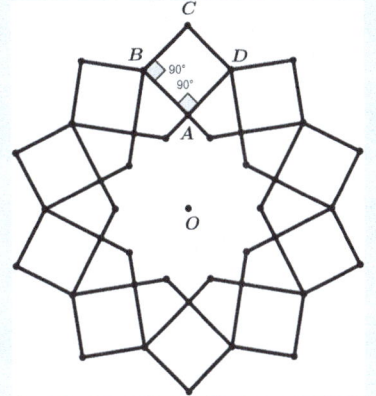
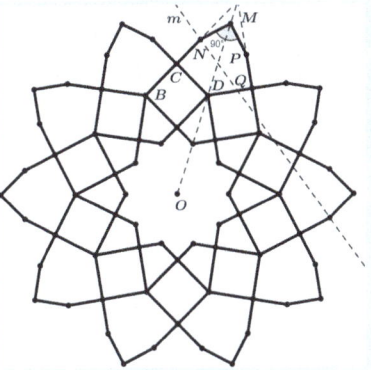

Drawing of the model 14. Step 1
1. Start from a (10,54°)2 = |10/3.5 |2 regular star. Draw: The point C rotation of the point A of center the point B and angle of 90°. Draw the square ABCD. Draw the rest of the squares with successive rotations of the square ABCD with center O and angle of 36°.

Drawing of the model 14. Step 2
2. Select the point M on the ray of origin O through D so that: N belongs to the perpendicular bisector m of C and M; the angle of vertex M is of 90°. Draw the rest of the hexagons with successive rotations of the hexagon DCNMPQ with center O and angle of 36°.

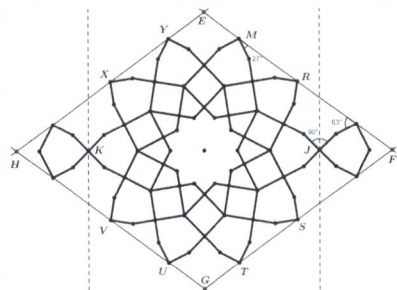
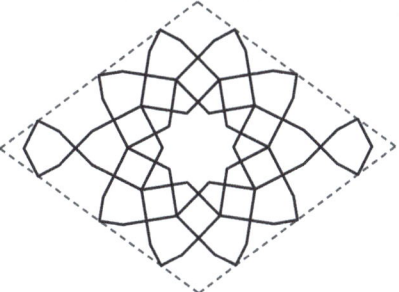

Drawing of the model 14. Step 3
3. Draw the symmetrical with respect to the vertical line through J of the hexagon located to its left and the symmetrical with respect to the vertical line through K of the hexagon located to its right.

Drawing of the model 14. Step 4
4. Repeat unit.

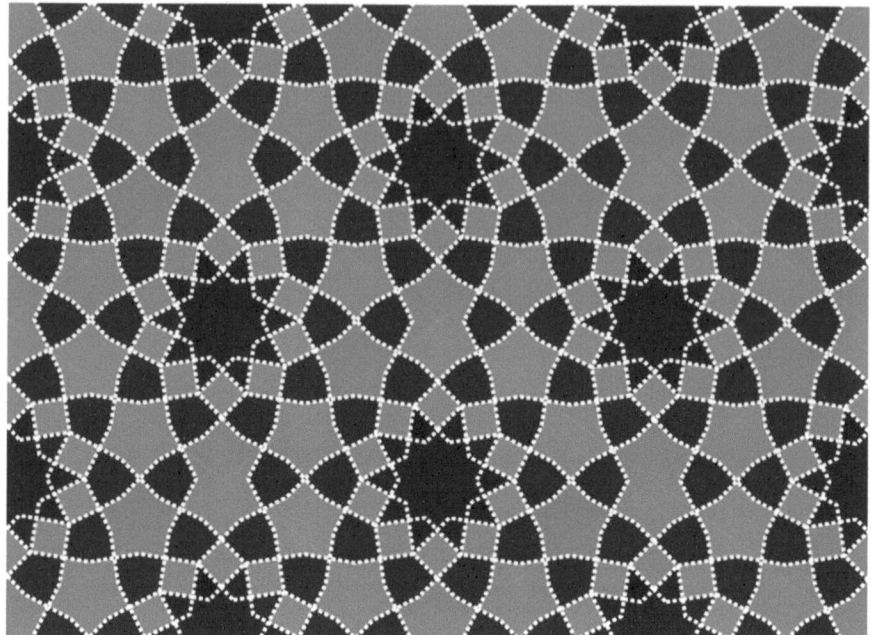

Fig. 4.28 Abdullah Khan Madrasa, Bukhara, Uzbekistan

Obtain the final periodic tessellation by multiple translations of the repeat unit (Fig. 4.28).

The tessellation contains $(10,54°)2 = |10/3.5|2$ and $(5,90°) = |5/1.25|$ regular stars.

4.1 Introduction

4.1.15 Fatima's Haram, Qom, Iran

One can construct the tessellation of Fig. 4.29 using a repeat unit which is a flattened hexagon.

The design of Fig. 4.29 contains convergent 10-pointed regular stars.

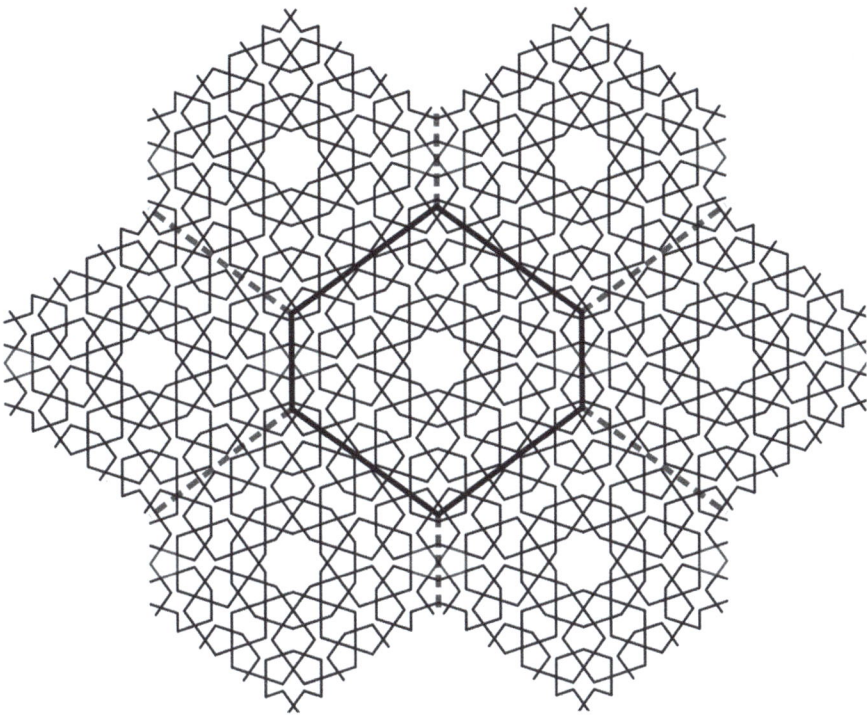

Fig. 4.29 Periodic tessellation structure of model 15

Construction of the repeat unit **Level: Difficult**

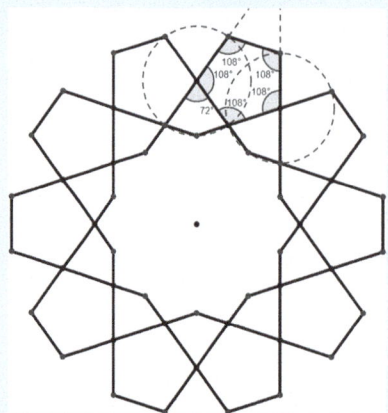

Drawing of the model 15. Step 1
1. Start from a regular star (10, 72°)2 = |10/3|2. Draw regular pentagons with interior angles of 108 degrees inserted into the dents of the star as shown.

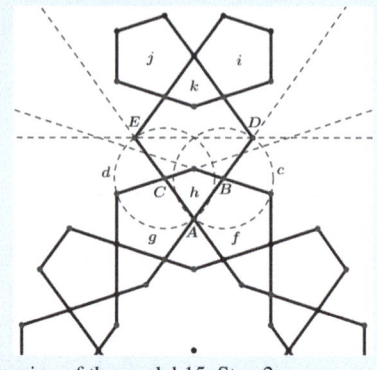

Drawing of the model 15. Step 2
2. Draw: The circle c of center B passing through A and the circle d of center C passing through A; the points D intersection of c and the ray of origin A passing through B and E intersection of d and the ray of origin A passing through C; the pentagons i and j and the triangle k symmetrical to pentagons f and g and triangle h with respect to the line passing through D and E.

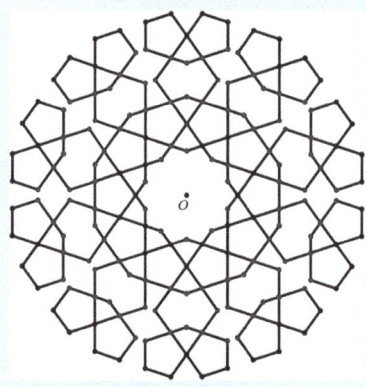

Drawing of the model 15. Step 3
3. Expand the drawing by successive rotations of center O and angle 36° of the part of the design constructed in step 2.

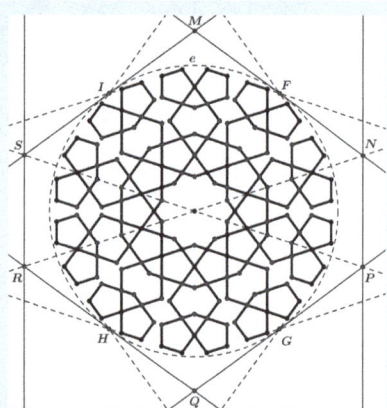

Drawing of the model 15. Step 4
4. Draw: The circle e passing through the points F, G, H, and I where intersect rays passing through the exterior sides of some outer pentagons as shown; the tangents to the circle e at F, G, H, and I; the polygon, boundary of the repeat unit, of vertices M, N, P, Q, R, and S as shown.

4.1 Introduction

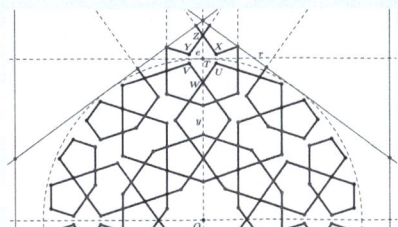

Drawing of the model 15. Step 5
5. Draw: The point T intersection of the circle e with the vertical line y passing through O; the tangent r to e at T; the points X, Y, and Z symmetrical of the points U, V, and W with respect to r. complete the drawing of the top of the interstitial region as shown.

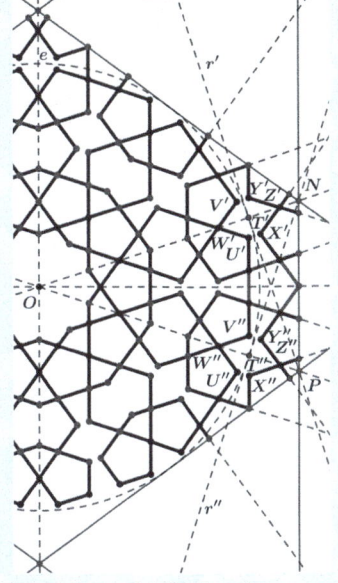

Drawing of the model 15. Step 6
6. Draw: The point T′ intersection of the circle e with the line r′ passing through O and N; the tangent r′ to e at T′; the points X′, Y′, and Z′ symmetrical of the points U′, V′, and W′ with respect to r′; the point T″ intersection of e with the line r″ passing through O and P; the tangent r″ to e at T″; the points X″, Y″, and Z″ symmetrical of the points U″, V″, and W″ with respect to r″. complete the drawing of the right side of the interstitial region as shown.

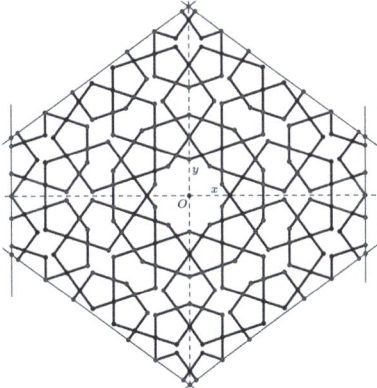

Drawing of the model 15. Step 7
7. Complete the drawing within the interstitial region using axial symmetries with respect to the lines x and y.

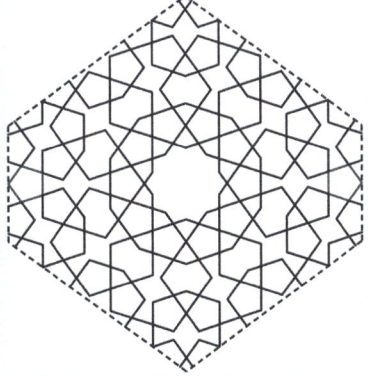

Drawing of the model 15. Step 8
8. Repeat unit

Fig. 4.30 Fatima's Haram, Qom, Iran

Obtain the final periodic tessellation by multiple translations of the repeat unit (Fig. 4.30).

4.1 Introduction

4.1.16 Masjid-i-Jami, Varamin, Iran

One can construct the tessellation of Fig. 4.31 using a repeat unit which is a regular hexagon.

The design of Fig. 4.31 contains convergent 12-pointed regular stars.

Fig. 4.31 Periodic tessellation structure of model 16

Construction of the repeat unit **Level: Easy**

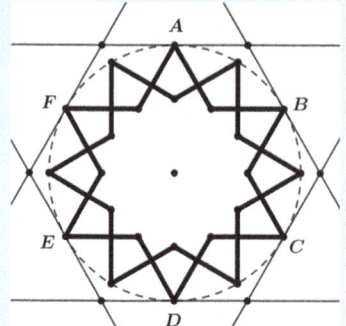 

Drawing of the model 16. Step 1
1. Start from a convergent $(12,60°)2 = |12/4|2$ regular star. Draw the circle c that passes through the spike vertices of the star. The boundary polygon is the regular hexagon bounded by the tangents to the circle c at the spike vertices A, B, C, D, E, and F of the star.

Drawing of the model 16. Step 2
2. The repeat unit

4.1 Introduction

Fig. 4.32 Masjid-i-Jami, Varamin, Iran

Obtain the final periodic tessellation by multiple translations of the repeat unit (Fig. 4.32).

4.1.17 Jameh Mosque, Kerman, Iran

One can construct the tessellation of Fig. 4.33 using a repeat unit which is a regular hexagon.

The design of Fig. 4.33 contains convergent 12-pointed and parallel 6-pointed regular stars.

Fig. 4.33 Periodic tessellation structure of model 17

4.1 Introduction

Construction of the repeat unit **Level: Medium**

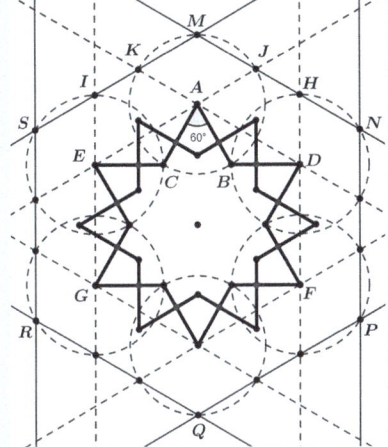

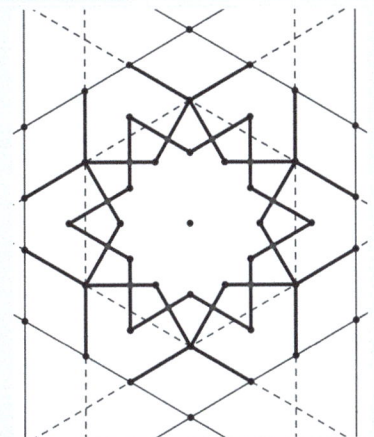

Drawing of the model 17. Step 1
1. Start from a (12,60°)2 = |12/4|2 regular star. Draw: The point J intersection of the circle with center the point A passing through the point B and the line passing through the points A and E; the point of intersection K of the circle with center A passing through the point C and the line passing through the points A and D; the point H intersection of the circle with center D passing through B and the line passing through the points F and D; the point I intersection of the circle with center the point E passing through C and the line passing through the points G and E; the point M intersection of the lines passing through the points H and J and through I and K. Draw in the same way the points N, P, Q, R, and S that together with M determine the boundary of the hexagonal repeated unit.

Drawing of the model 17. Step 2
2. Draw part of the model inside the interstitial region as shown.

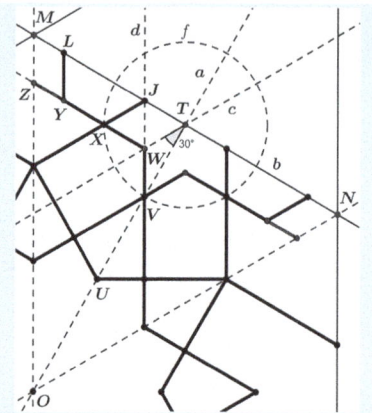
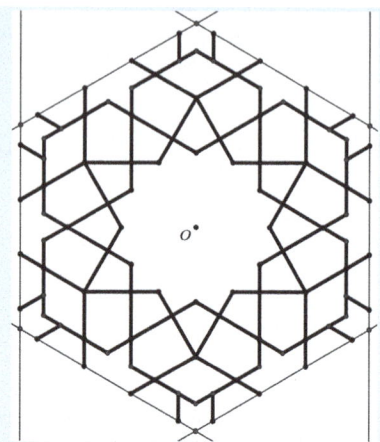

Drawing of the model 17. Step 3

3. Draw: The line a through points O and U; the point T intersection of a with line b passing through points M and N; the line c rotated of a with center T and angle of 30° clockwise; the ray d of origin V through J; the point W intersection of c and d; the point X rotated of V with center T and 60° clockwise; the point Y symmetrical of W with respect to X; the point L symmetrical of V with respect to X; the point Z intersection of the vertical through O and the ray of origin X through Y. Draw the design inside the triangle of vertices O, T, and M as shown. Obtain the corresponding design inside the triangle of vertices O, N, and T by a symmetry with respect to the line a.

Drawing of the model 17. Step 4

4. Complete the drawing within the interstitial region by successive rotations of center O and angle of 60° of the triangular upper right part drawn in the step 3.

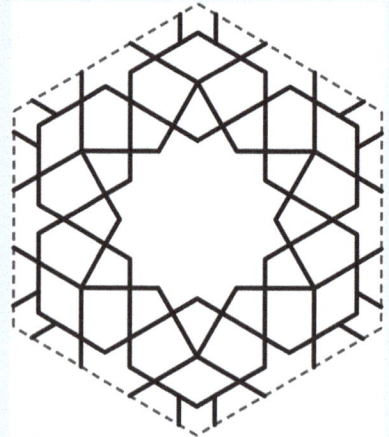

Drawing of the model 17. Step 5

5. Repeat unit.

4.1 Introduction

Fig. 4.34 Jameh Mosque, Kerman, Iran

Obtain the final periodic tessellation by multiple translations of the repeat unit (Fig. 4.34).

The tessellation contains convergent $(12,60°)2 = |12/4|2$ and two kinds of parallel $(6,60°) = |6,2|$ regular stars (green and brown).

Note that the white lines represent the pattern of Model 3. Tash Hauli Palace Complex, Khiva, Uzbekistan.

4.1.18 Hall of Kings, Alhambra, Granada, Spain

One can construct the tessellation of Fig. 4.35 using a repeat unit which is a regular hexagon.

The design of Fig. 4.35 contains parallel 12-pointed regular stars.

Fig. 4.35 Periodic tessellation structure of model 18

4.1 Introduction

Construction of the repeat unit **Level: Medium**

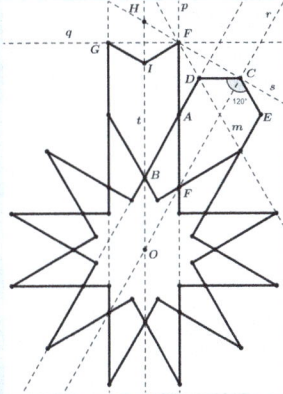

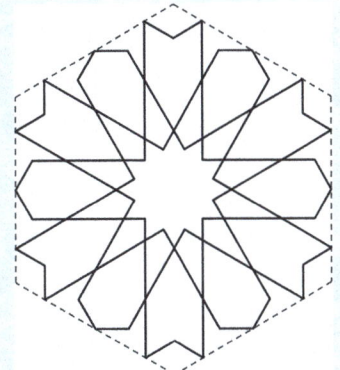

Drawing of the model 18. Step 1

1. Start from a parallel $(12,30°)2 = |12/5|2$ regular star. Draw: The line r through points O and F; the point C on r so that: D belongs to the perpendicular bisector m of A and C and the angle of vertex C is of 120°; the point E symmetric of the point D with respect to line r; the vertical p through the point A; the line t through the points O and B; the point F intersection of the lines m and p; the point G symmetric of the point F with respect to the line t; the line q through the points F and G; the line s through the points C and F; the point H intersection of the lines s and t; the point I symmetric of the point H with respect to the line q. Draw a part of the design as shown.

Drawing of the model 18. Step 3
3. Repeat unit.

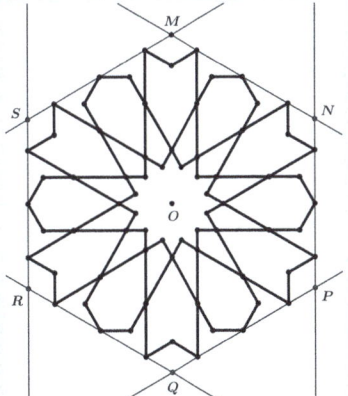

Drawing of the model 18. Step 2
2. Draw the rest of the interstitial region with successive rotations of center O and angle of 30° of the part drawn in paragraph 1. The regular hexagon of vertices M, N, P, Q, R, and S is the boundary of the repeat unit.

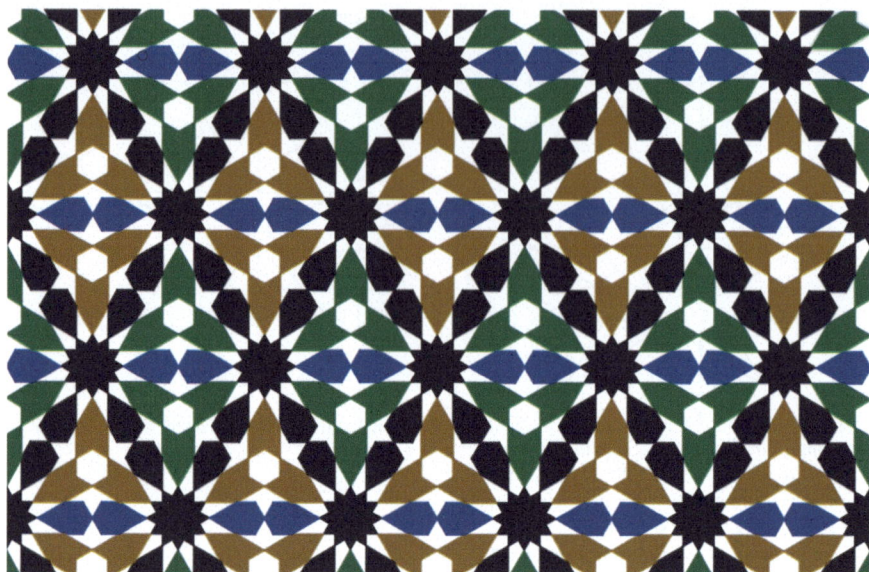

Fig. 4.36 Hall of Kings, Alhambra, Granada, Spain

Obtain the final periodic tessellation by multiple translations of the repeat unit (Fig. 4.36).

4.1 Introduction

4.1.19 Bibi Khanum Mosque, Samarkand, Uzbekistan

One can construct the tessellation of Fig. 4.37 using a repeat unit which is a square.

The design of Fig. 4.37 contains convergent 12-pointed and 8-pointed regular stars.

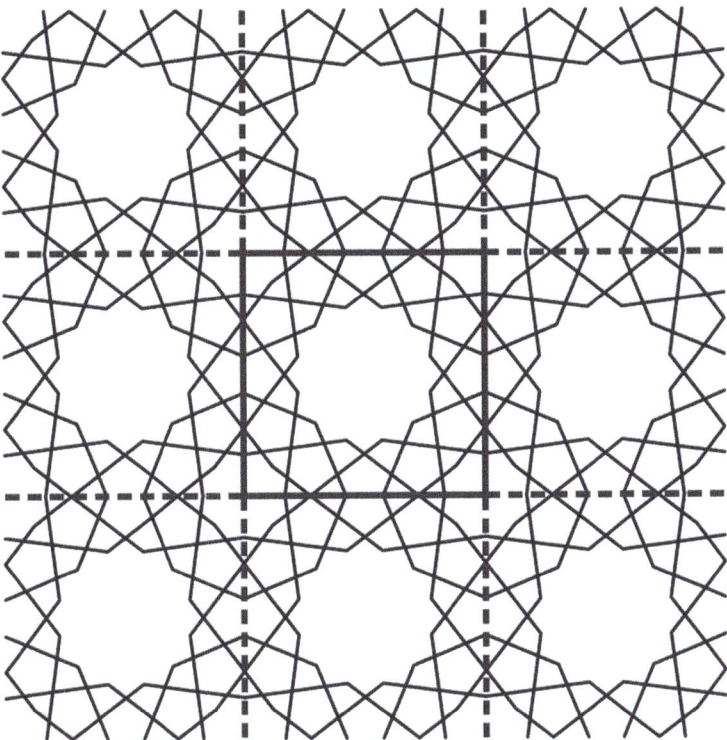

Fig. 4.37 Periodic tessellation structure of model 19

Construction of the repeat unit   **Level: Medium**

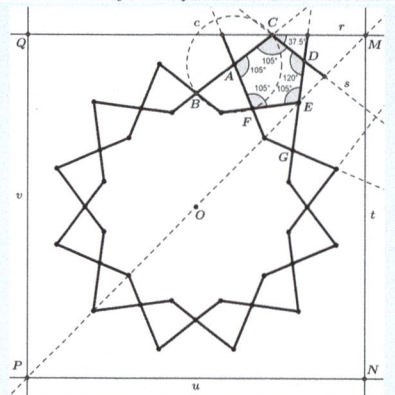

Drawing of the model 19. Step 1

1. Start from a convergent $(12,75°)2 = |12/3.5|2$ regular star. Draw: The circle c of center the point A through the point B; the point C intersection of the circle c with the ray of origin the point B passing through point A; the horizontal line r through point C; the line s rotation of the line r of center the point C and angle of 37.5° clockwise; the point D intersection of the line s with the ray of origin the point G passing through E; the lines t, u, and v obtained from the line r with succesive rotations of center the point O and 90°. The lines r, t, u, and v determine the square of vertices M, N, P, and Q which is the boundary of the repeat unit. Complete the drawing as shown.

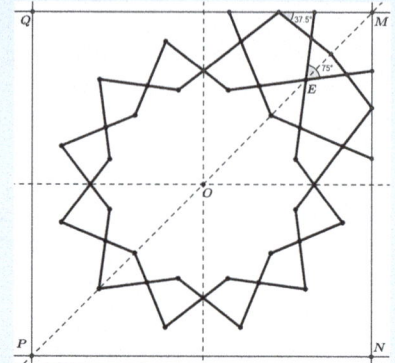

Drawing of the model 19. Step 2

2. Complete the upper right part of the design using a symmetry of axis the line through O and E.

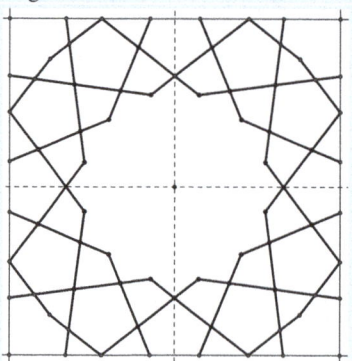

Drawing of the model 19. Step 3

3. Complete the drawing within the interstitial region of the repeat unit using axial symmetries.

Drawing of the model 19. Step 4

4. Repeat unit.

4.1 Introduction

Fig. 4.38 Bibi Khanum Mosque, Samarkand, Uzbekistan

Obtain the final periodic tessellation by multiple translations of the repeat unit (Fig. 4.38).

The tessellation contains convergent (12,75°)2 = |12/3.5|2 and (8,75°)2 = |8/(7/3)|2 regular stars.

4.1.20 Qalawun Mausoleum, Cairo, Egypt

One can construct the tessellation of Fig. 4.39 using a repeat unit which is a regular hexagon.

The design of Fig. 4.39 contains convergent 12-pointed, parallel 6-pointed, and divergent 4-pointed regular stars.

Fig. 4.39 Periodic tessellation structure of model 20

4.1 Introduction

Construction of the repeat unit

Level: Medium

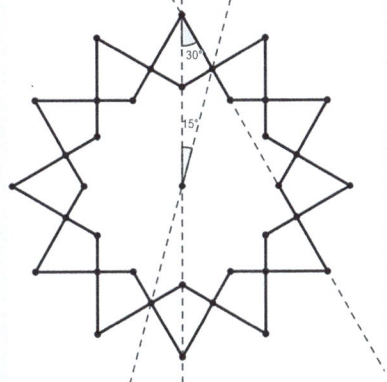

Drawing of the model 20. Step 1
1. Start from a regular star (12,60°)2 = |12/4|2.

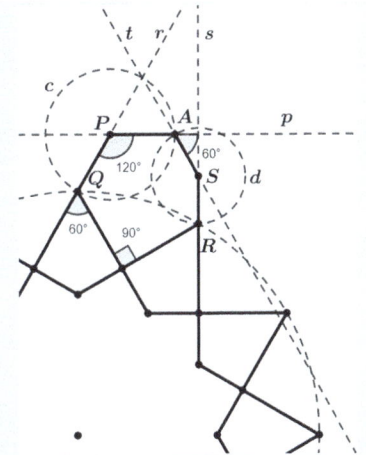

Drawing of the model 20. Step 2
2. Construct an asymmetrical petal as follows. Draw: a point P draggable along the line r, which we will determine later; the line p rotation of r of center P and 120°; the circle c of center P through Q; the point A intersection of p and c; the line t rotation of p with center the point A and 60°; the point S intersection of the lines s and t. choose the point P on the line r so that the circle d with center the point S passing through A also passes through R.

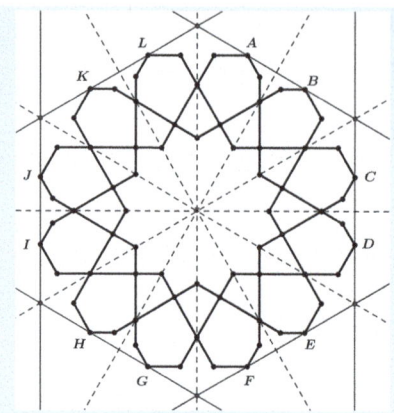

Drawing of the model 20. Step 3
3. Complete the drawing with successive axial symmetries of the asymmetrical petal. Lines passing through points A and B, C and D, E and F, G and H, I and J, K and L determine the boundary of the repeating unit.

Drawing of the model 20. Step 4
4. Repeat unit.

4.1 Introduction

Fig. 4.40 Qalawun Mausoleum, Cairo, Egypt

Obtain the final periodic tessellation by multiple translations of the repeat unit (Fig. 4.40).

The tessellation contains convergent $(12,60°)2 = |12/4|2$, parallel $(6,60°) = |6/2|$, and divergent $(4,60°) = |4/(4/3)|$ regular stars.

4.1.21 Hot Room, Bath of Comares, Alhambra, Granada, Spain

One can construct the tessellation of Fig. 4.41 using a repeat unit which is a square. The design of Fig. 4.41 contains divergent 4-pointed regular stars.

Fig. 4.41 Periodic tessellation structure of model 21

4.1 Introduction

Construction of the repeat unit

Base unit and repeat unit
We will begin by drawing the base unit bounded by a square located to the right of the square repeat unit, and then complete it using axial symmetries.

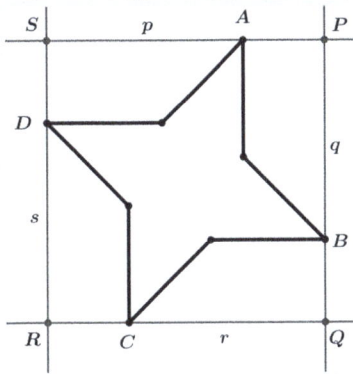

Drawing of the model 21. Step 2
2. Draw: The horizontal lines p passing through point A and r passing through point C; the vertical lines q passing through point B and s passing through point D. The vertices of the boundary of the base unit are the points P, Q, R, and S where intersect the lines p and q, q and r, r and s, and s and p, respectively.

Level: Easy

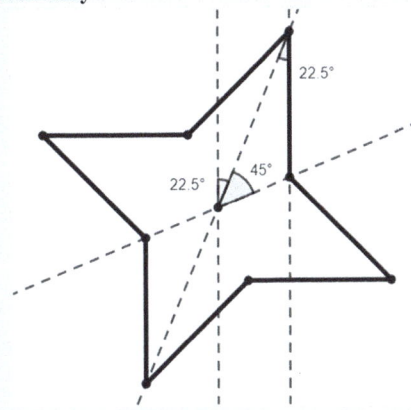

Drawing of the model 21. Step 1
1. Draw a (4,45°) = |4/1.5| regular star.

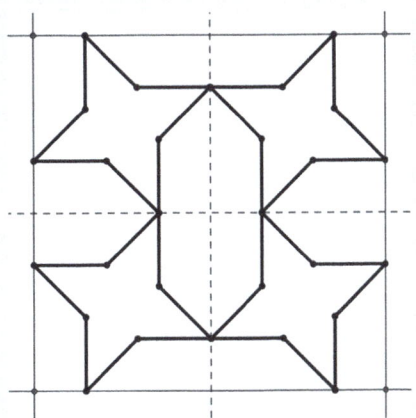

Drawing of the model 21. Step 3
3. Complete the repeat unit using axial symmetries.

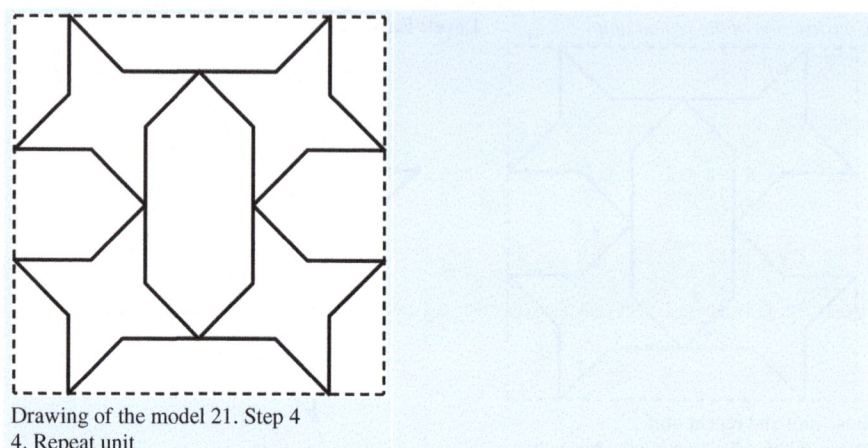

Drawing of the model 21. Step 4
4. Repeat unit

4.1 Introduction

Fig. 4.42 Hot Room, Bath of Comares, Alhambra, Granada, Spain

Obtain the final periodic tessellation by multiple translations of the repeat unit (Fig. 4.42).

4.1.22 Mashhad al-Imam Yahya ibn al-Qasim, Mosul, Iraq

One can construct the tessellation of Fig. 4.43 using a repeat unit which is a flattened hexagon.

The design of Fig. 4.43 contains convergent 7-pointed regular stars.

Fig. 4.43 Periodic tessellation structure of model 22

4.1 Introduction

Construction of the repeat unit

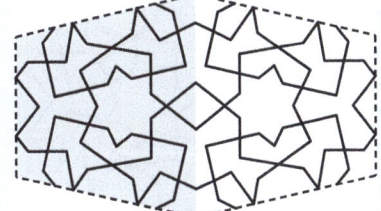

Base unit and repeat unit
We will begin by drawing the base unit bounded by an isosceles trapezoid located to the right of the hexagonal repeat unit, and then complete it using an axial symmetry.

Level: Difficult

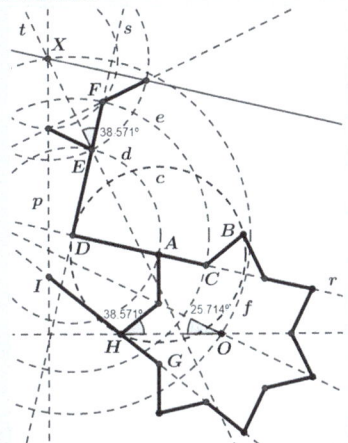

Drawing of the model 22. Step 1
1. Start from a convergent regular star $(7, 77.143°) = |7/2|$. Draw: The circle c of center A through B; the line r through A and C; the points D intersection of r and c; the line s orthogonal to r through D; the circles d and e of center D passing through A and C, respectively; the points E and F intersection of s with d and e, respectively; the line t rotated of s by a rotation of of center E and angle of 38.571° counterclockwise; the point X intersection of t with the circle of center D through O; the point I intersection of the line through G and H with the vertical p through X. Complete the drawing as shown.

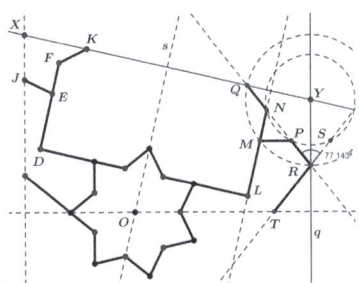

Drawing of the model 22. Step 2
2. Draw: The points Y, L, M, N, P, and Q symmetrical with respect to the line s of points X, D, E, F, J, and K, respectively; the vertical q through Y; the point S symmetrical of P with respect to q; the point R intersection of the line through Q and N with q; the point T intersection of the line through S and R with the horizontal through O. Complete the drawing as shown.

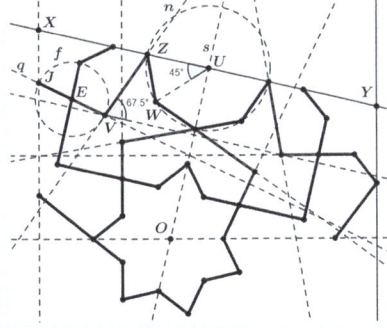

Drawing of the model 22. Step 3
3. Draw: The circle f of center E passing through J; the line q passing through J and E; the point V intersection of q and f; the line n rotated of q by a rotation of center V and angle 67.5° counterclockwise; the point Z intersection of n with the line passing through X and Y; the point W rotated of Z by a rotation of center U and angle of 45°. Complete the drawing as shown.

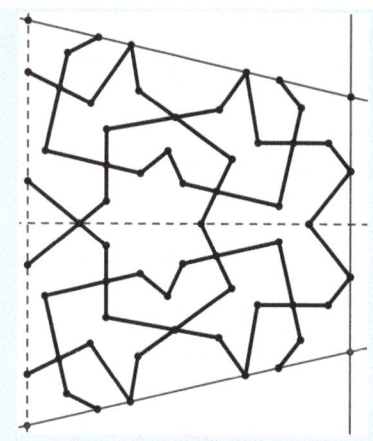

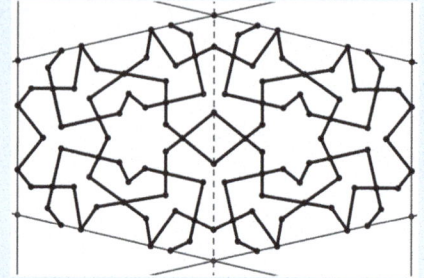

Drawing of the model 22. Step 5
5. Obtain the repeat unit from the base unit using an axial symmetry.

Drawing of the model 22. Step 4
4. Draw the bottom of the base unit with an axial symmetry of the top as shown.

4.1 Introduction

Fig. 4.44 Mashhad al-Imam Yahya ibn al-Qasim, Mosul, Iraq

Obtain the final periodic tessellation by multiple translations of the repeat unit (Fig. 4.44).

4.1.23 Tomb of I'timād-ud-Daulah, Agra, India

One can construct the tessellation of Fig. 4.45 using a repeat unit which is a square. The design of Fig. 4.45 contains 8-pointed convergent regular stars.

Fig. 4.45 Periodic tessellation structure of model 23

4.1 Introduction

Construction of the repeat unit   **Level: Medium**

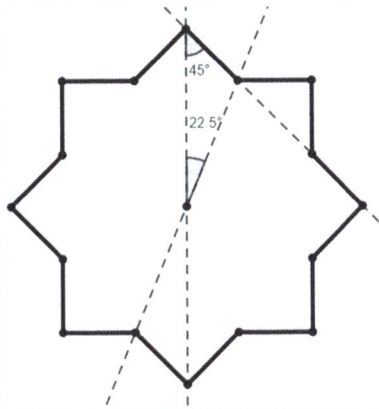

Base unit and repeat unit
We will start by drawing the triangular base unit located at the upper right quadrant of the square repeat unit, and then complete it using axial symmetries.

Drawing of the model 23. Step 1
1. Start from a regular star $(8, 90°) = |8/2|$.

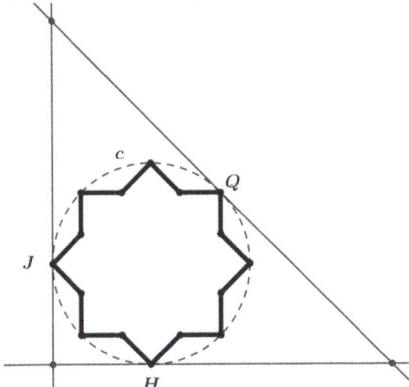

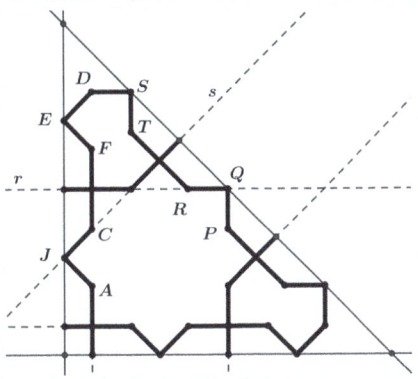

Drawing of the model 23. Step 2
2. Draw the circle c that passes through the spike vertices of the star. The boundary polygon f of the base unit is the triangle bounded by the tangents to the circle c at the spike vertices J, H, and Q of the star.

Drawing of the model 23. Step 3
3. Draw within the upper part of the interstitial region of the base unit the points: D, E, and F symmetrical of A, J, and C with respect to the line r; points S and T symmetrical of Q and R with respect to the line s. Draw the equivalent points within the lower right part of the interstitial region of the base unit. Complete the drawing of the base unit as shown.

Drawing of the model 23. Step 4
4. Complete the repeat unit from the base unit using axial symmetries.

4.1 Introduction

Fig. 4.46 Tomb of I'timād-ud-Daulah, Agra, India

Obtain the final periodic tessellation by multiple translations of the repeat unit (Fig. 4.46).

4.1.24 Mosaic of the Alhambra Museum, Granada, Spain

One can construct the tessellation of Fig. 4.47 using a repeating unit which is a square.

The design of Fig. 4.47 contains parallel 8-pointed regular stars.

Fig. 4.47 Periodic tessellation structure of model 24

4.1 Introduction

Construction of the repeat unit **Level: Easy**

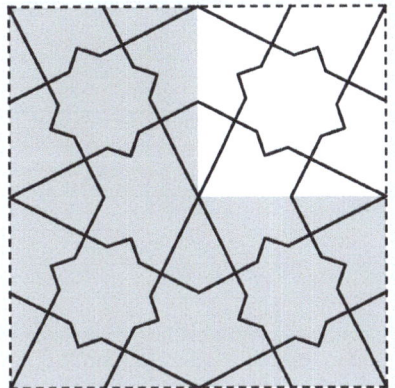

Base unit and repeat unit
We will start by drawing the unit base of the upper right quadrant of the quadrangular repeat unit and then complete it using axial symmetries.
Since the star of the basic unit does not have a vertical axis of symmetry, we will have to find the inclination of an axis of symmetry from which to start drawing the star.

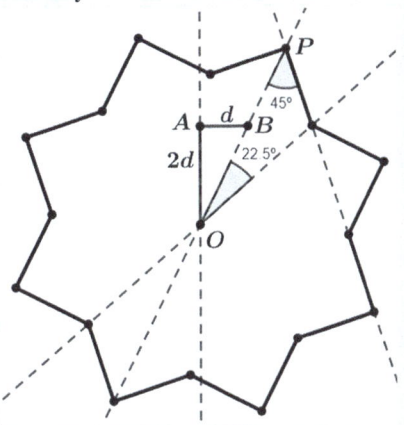

Drawing of the model 24. Step 1
Choose point A on the vertical through the point that will be the center O of the star and point B so that the triangle OAB is right-angled at A and the length of the side OA is twice that of AB. Choose the initial vertex P of the star on the line passing through O and B. Draw the regular star $(8, 90°) = |8, 2|$ with center O and initial vertex P.

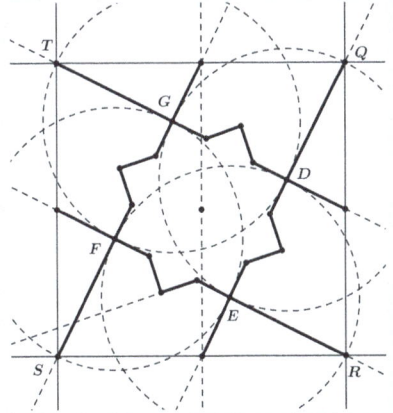

Drawing of the model 24. Step 2
Draw the circles with center the point D and radius DE, center the point E and radius EF, center the point F and radius FG, and center the point G and radius GD. Draw the points Q, R, S, and T of intersection of the circles with the extensions of the sides of the star as shown. The boundary polygon f of the base unit is the square of vertices Q, R, S, and T. Extend the sides of the vertices that do not belong to the boundary f and find the points where they intersect it. Then complete the drawing within the interstitial region of the base unit as shown.

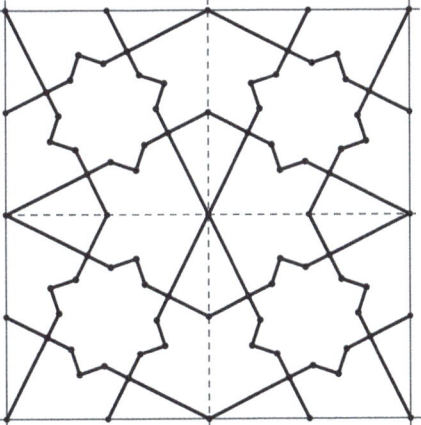

Drawing of the model 24. Step 3
Complete the repeat unit from the base unit using axial symmetries.

Drawing of the model 24. Step 4 Repeat unit.

4.1 Introduction

Fig. 4.48 Mosaic of the Alhambra Museum, Granada, Spain

Obtain the final periodic tessellation by multiple translations of the repeat unit (Fig. 4.48).

4.1.25 Ulugh Beg Madrasa, Samarkand, Uzbekistan

One can construct the tessellation of Fig. 4.49 using a repeat unit which is a rhombus.

The design of Fig. 4.49 contains convergent 9-pointed and 6-pointed regular stars.

Fig. 4.49 Periodic tessellation structure of model 25

4.1 Introduction

Construction of the repeat unit

Level: Medium

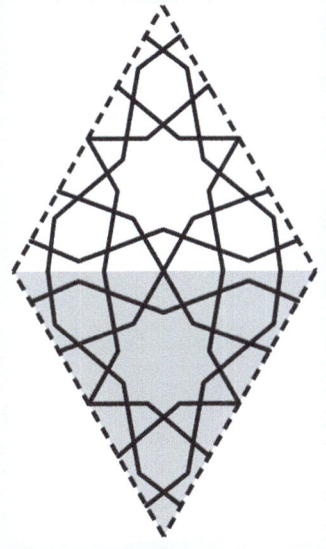

Base unit and repeat unit
We will draw the base unit bounded by an equilateral triangle on top of the rhombic repeat unit, and then complete it with an axial symmetry.

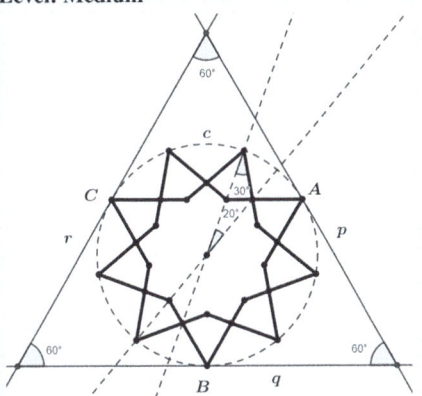

Drawing of the model 25. Step 1
1. Start from a (9,60°)2 = |9/3|2 regular star. Draw: The circle c that passes through the spike vertices of the star. The boundary polygon f of the base unit is the isosceles triangle bounded by the tangents p, q, and r to the circle c at the spike vertices A, B, and C.

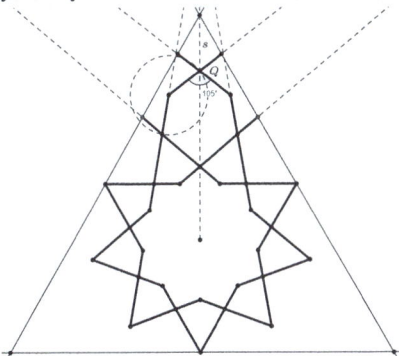

Drawing of the model 25. Step 2
2. Draw on top of the star a standard petal by selecting point Q on line s so that the interior angle at point Q measures 105°. Complete the drawing of the upper part of the interstitial region by extending the sides of the vertices that do not belong to the boundary f and finding the points where they intersect it.

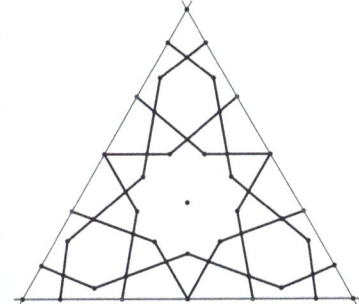

Drawing of the model 25. Step 3
3. Complete the drawing within the interstitial region of the base unit with successive rotations of center O and angle of 60° of the petal and its extensions previously drawn on top of the interstitial region.

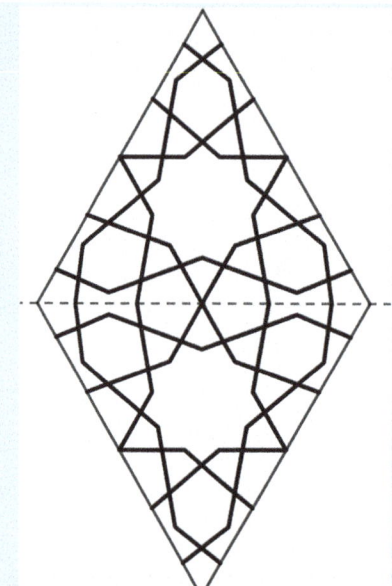

Drawing of the model 25. Step 4
4. Obtain the repeat unit from the base unit using an axial symmetry.

4.1 Introduction

Fig. 4.50 Ulugh Beg Madrasa, Samarkand, Uzbekistan

Obtain the final periodic tessellation by multiple translations of the repeat unit (Fig. 4.50).

The tessellation contains convergent $(9,60°)2 = |9/3|2$ and $(6105°) = |6/1.25|$ regular stars.

4.1.26 Ahmad ibn Tulun Mosque, Cairo, Egypt

One can construct the tessellation of Fig. 4.51 using a repeat unit which is a regular hexagon.

The design of Fig. 4.51 contains parallel ideal 6-pointed rosettes.

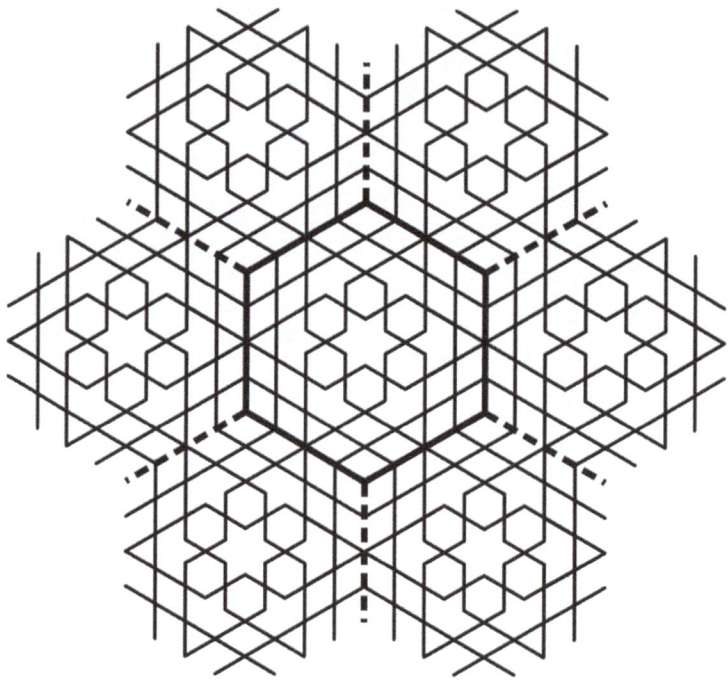

Fig. 4.51 Periodic tessellation structure of model 26

4.1 Introduction

Construction of the repeat unit   **Level: Easy**

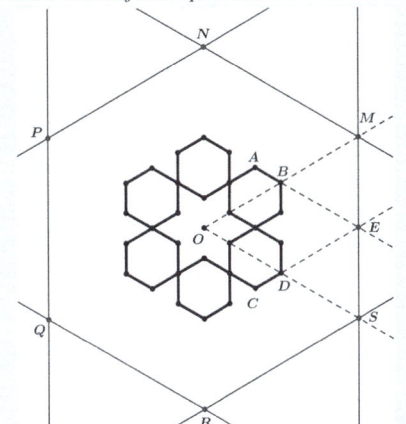

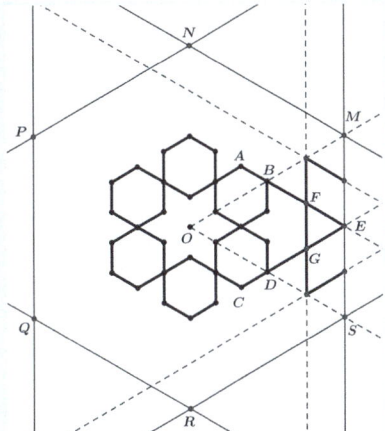

Drawing of the model 26. Step 1
1. Start from a parallel ideal 6-pointed rosette with a central regular star $(6,60°) = |6/2|$. Draw: The point E intersection of the ray of origin the point A passing through the point B and the ray of origin the point C passing through the point D; the point M intersection of the vertical through the point E and the ray of origin the point O through the point B; the point S intersection of the vertical through the point E and the ray of origin O through the point D. Draw the remaining vertices N, P, Q, and R of the boundary polygon of the repeat unit by successive rotations of center O and angle of 60°.

Drawing of the model 26. Step 2
2. Draw: The midpoint F of the points B and E; the midpoint G of the points D and E. Complete the drawing within the right side of the interstitial region as shown.

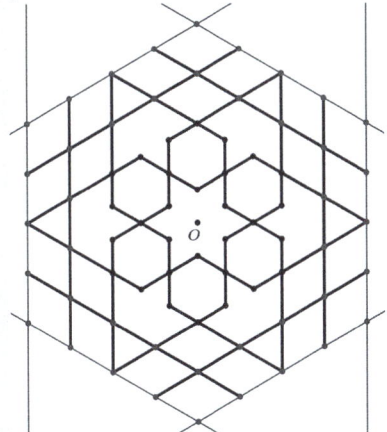

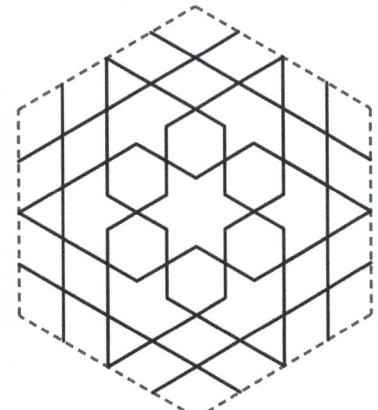

Drawing of the model 26. Step 3
3. Complete the drawing within the interstitial region by successive rotations of center O and angle of 60° of the triangular right part drawn in step 3.

Drawing of the model 26. Step 4
4. Repeat unit.

Fig. 4.52 Ahmad ibn Tulun Mosque, Cairo, Egypt

Obtain the final periodic tessellation by multiple translations of the repeat unit (Fig. 4.52).

4.1.27 Mazar of Sachal Sarmast, Khairpur, Pakistan

One can construct the tessellation of Fig. 4.53 using a repeat unit which is a square. The design of Fig. 4.53 contains 8-pointed parallel ideal rosettes.

Fig. 4.53 Periodic tessellation structure of model 27

Construction of the repeat unit **Level: Easy**

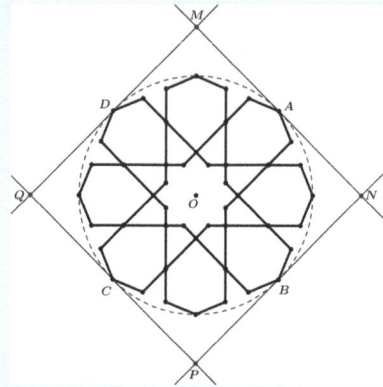

Drawing of the model 27. Step 1
1. Start from a parallel ideal 8-pointed rosette with a central regular star $(8,45°)2 = |8/3|2$. Draw the tangents to the circle through the tips of the petals at points A, B, C, and D. The points M, N, P, and Q of intersection of the tangents are the vertices of the squared boundary of the repeat unit.

Drawing of the model 27. Step 2
2. Extend the sides of the tip petals at points X, Y, Z, and W and intersect them with the repeat unit boundary.

Drawing of the model 27. Step 3
3. Complete the drawing within the interstitial region of the repeat unit.

Drawing of the model 27. Step 4
4. Repeat unit.

4.1 Introduction 133

Fig. 4.54 Mazar of Sachal Sarmast, Khairpur, Pakistan

Obtain the final periodic tessellation by multiple translations of the repeat unit (Fig. 4.54).

4.1.28 Hudavent tomb, Nigde, Turkey

One can construct the tessellation of Fig. 4.55 using a repeat unit which is an elongated hexagon.

The design of Fig. 4.55 contains 8-pointed ideal parallel rosettes.

Fig. 4.55 Periodic tessellation structure of model 28

4.1 Introduction

Construction of the repeat unit **Level: Medium**

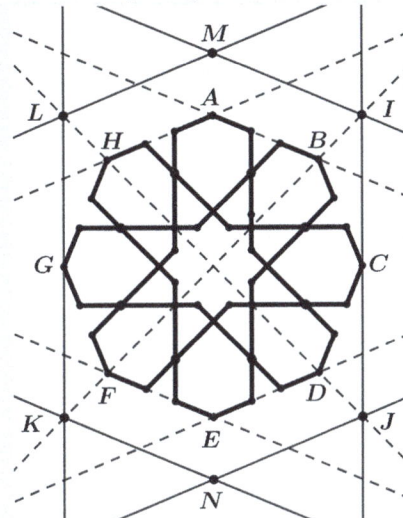

Drawing of the model 28. Step 1
1. Start from a parallel ideal 8-point rosette with a central regular star $(8,45°)2 = |8/3|2$. The boundary polygon f of the repeat unit is the elongated hexagon of vertices: The point of intersection I of the vertical line through C with the line through B and F; the intersection point J of the vertical line through C with the line through D and H; the intersection point K of the vertical line through G with the line through B and F; the intersection point L of the vertical line through G with the line through D and H; the intersection point M of the line through I parallel to the line through A and B with the line through L parallel to the line through A and H; the intersection point N of the line through J parallel to the line through D and E with the line through K parallel to the line through E and F.

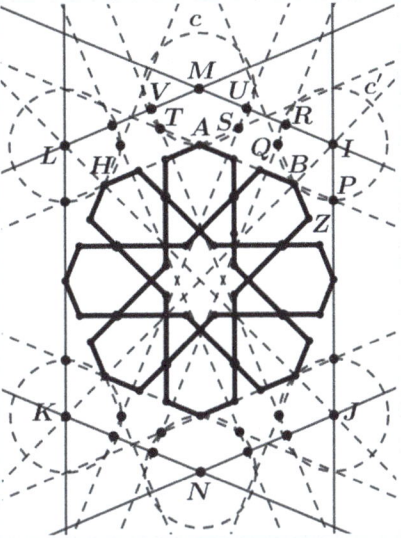

Drawing of the model 28. Step 2
2. We proceed to complete the interstitial region. We start with the part close to M. Draw: The circle c with center the point M that passes through the point A; the point S intersection of c with the line through the points H and A; the point T intersection of c with the line through the points B and A; the point U intersection of the line passing through the points I and M with the perpendicular to it through the point S; the point V intersection of the line through the points L and M with the perpendicular to it through the point T. The part of the interstitial region close to the point N is completed in the same way as we have done for the one close to the point M. To complete the part of the interstitial region close to the point I, first we draw the circle c' with center I that passes through the point B. We draw the point P intersection of c' with the line through A and B; the point Q intersection of c' with the line through the points Z and B; the point R intersection of the line through the points I and M with the perpendicular to it through the point Q. The parts of the interstitial region close to the points J, K, and L are completed in the same way as we have done for the one close to I.

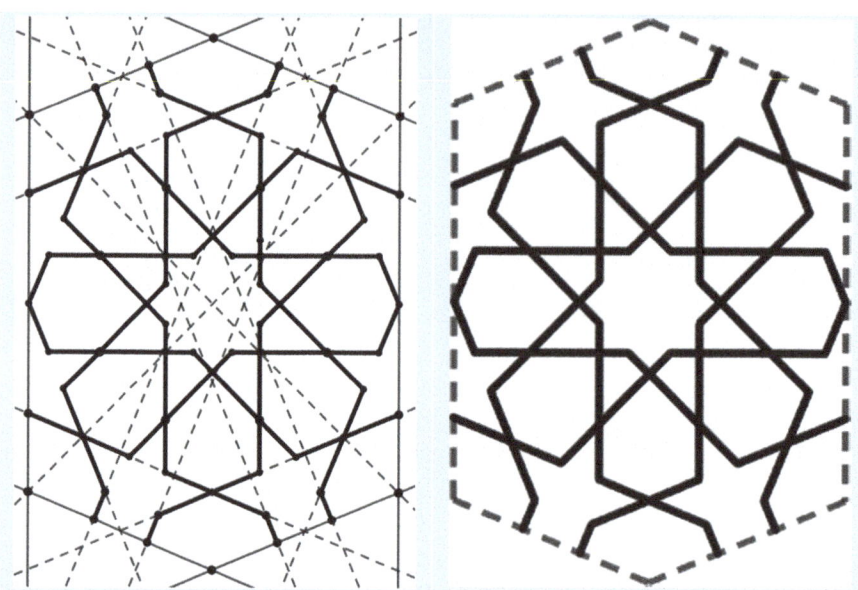

Drawing of the model 28. Step 3
3. Complete the drawing within the interstitial region of the repeat unit as shown.

Drawing of the model 28. Step 4
4. Repeat unit.

4.1 Introduction

Fig. 4.56 Hudavent tomb, Nigde, Turkey

Obtain the final periodic tessellation by multiple translations of the repeat unit (Fig. 4.56).

4.1.29 Alhambra, Granada, Spain

One can construct the tessellation of Fig. 4.57 using a repeat unit which is a square.

The design of Fig. 4.57 contains ideal parallel 8-pointed rosettes and convergent 8-pointed regular stars.

Fig. 4.57 Periodic tessellation structure of model 29

4.1 Introduction

Construction of the repeat unit **Level: Medium**

The width of the petals of the 8-pointed rosette is the same as the width of the figures just outside the 8-pointed rosette. Thus, the 8-pointed star is highly constrained.

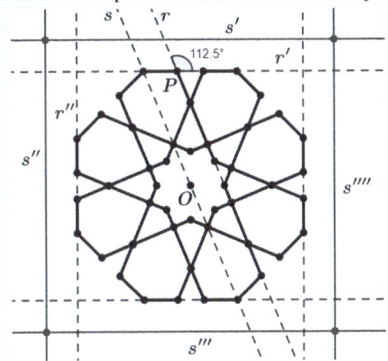

Drawing of the model 29. Step 1

1. Start from a parallel ideal rosette based on a $(8,45°)2 = |8/3|2$ regular star. Draw: The lines r' rotation of r of center P and $112.5°$ clockwise; the line s' rotation of the line s of center the point P and $112.5°$ clockwise; the line r'' rotation of r' of center the point O and $90°$ counterclockwise; the line s'' rotation of s' of center O and $90°$ counterclockwise; the line s''' rotation of s'' of center O and $90°$ counterclockwise; the line s'''' rotation of s''' of center O and $90°$ counterclockwise. The lines s', s'', s''', and s'''' determine the boundary polygon of the square repeat unit.

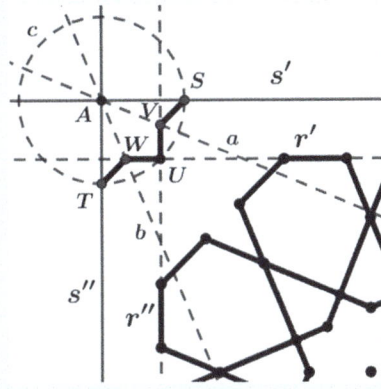

Drawing of the model 29. Step 2

2. The part of the $|8/2| = (8.90°)$ regular star of the upper left side of the repeat unit is determined by the lines r', s', r'', and s''. Draw: the point A intersection of the lines s' and s''; the point U intersection of the lines r' and r''; the circle c with center A passing through U; the points S and T intersection of c with s' and s''; the perpendicular bisectors a and b of the segments of endpoints S and U, and U and T; the points V and W intersection of lines a and b with lines r' and r''. The points S, V, U, W, and T are the vertices of the part of the $|8/2| = (8,90°)$ regular star of the upper left side of the repeat unit.

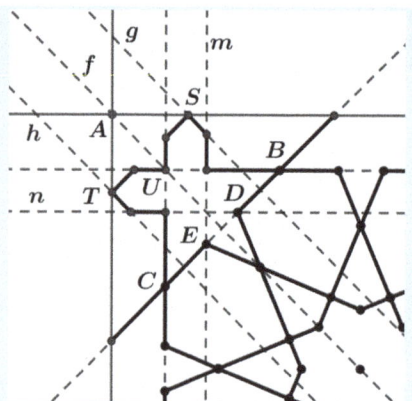
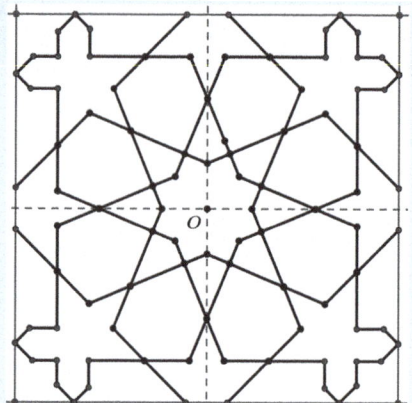

Drawing of the model 29. Step 3
3. Draw: the line f passing through A and U; the lines g and h parallel to f passing through the points S and T; the horizontal n through D; the vertical m through E; the extension of the sides of the petals of vertices B and C of the 8-pointed rosette. Complete the drawing in the upper left side of the repeat unit as shown.

Drawing of the model 29. Step 4
4. Complete the drawing within the interstitial region of the repeat unit by successive rotations of center O an angle of 90° of the upper left side of the repeat unit drawn in step 3.

4.1 Introduction

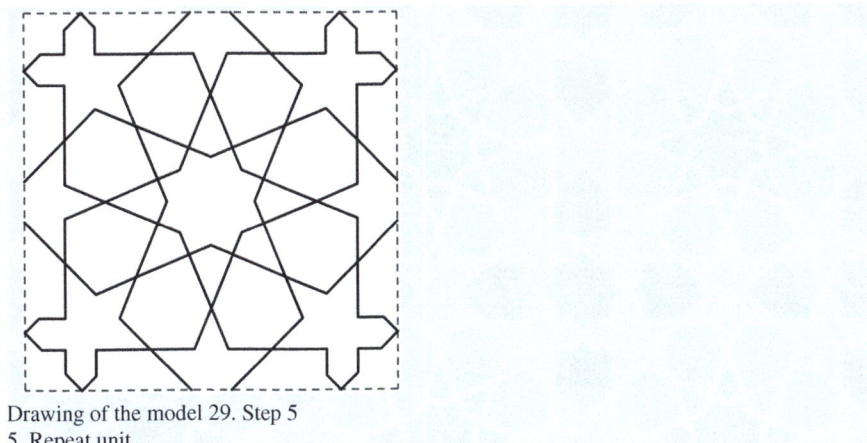

Drawing of the model 29. Step 5
5. Repeat unit.

Fig. 4.58 Alhambra, Granada, Spain

Obtain the final periodic tessellation by multiple translations of the repeat unit (Fig. 4.58).

The tessellation contains parallel ideal rosettes with center a (8,45°)2 = |8/3|2 regular star and (8,90°) = |8/2| regular stars.

4.1 Introduction

4.1.30 Kok Gumbaz Mosque, Shahrisabz, Uzbekistan

One can construct the tessellation of Fig. 4.59 using a repeat unit which is a rhombus.

The design of Fig. 4.59 contains convergent standard 10-pointed rosettes and parallel 5-pointed regular stars.

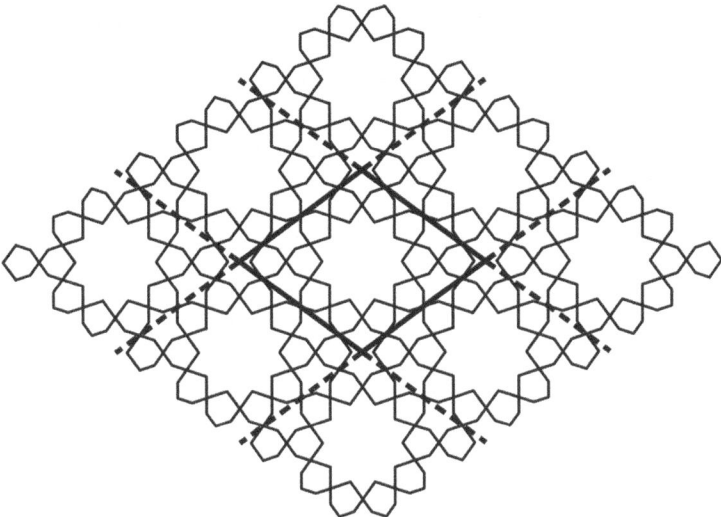

Fig. 4.59 Periodic tessellation structure of model 30

Construction of the repeat unit **Level: Medium**

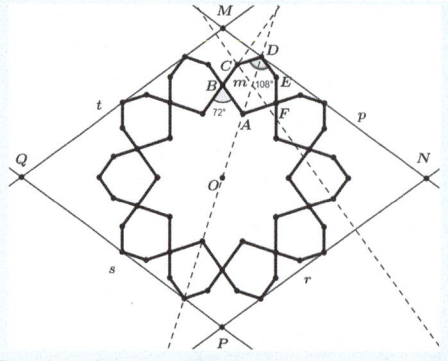

Drawing of the model 30. Step 1
1. Start from a convergent $(10, 72°) = |10/3|$ regular star. Draw the point D on the ray of origin O through A so that: C belongs to the perpendicular bisector m of B and D; the angle of vertex D is of 108°. Draw the rest of the petals of the rosette with successive rotations of the petal ABCDEF with center O and angle of 36°. Draw: the lines p, r, s, and t as shown. The rhombus of vertices M, N, P, and Q determines by the lines p, q, r, and s is the boundary of the repeat unit.

Drawing of the model 30. Step 2
2. Draw the symmetrical with respect to the vertical line through R of the petal located to its left and the symmetrical with respect to the vertical line through S of the petal located to its right.

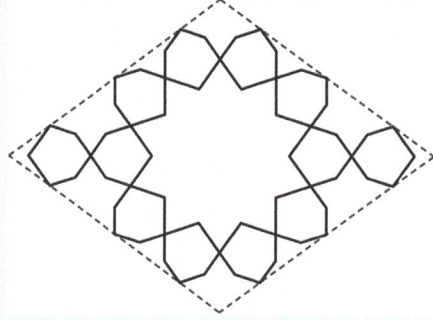

Drawing of the model 30. Step 3
3. Repeat unit

4.1 Introduction

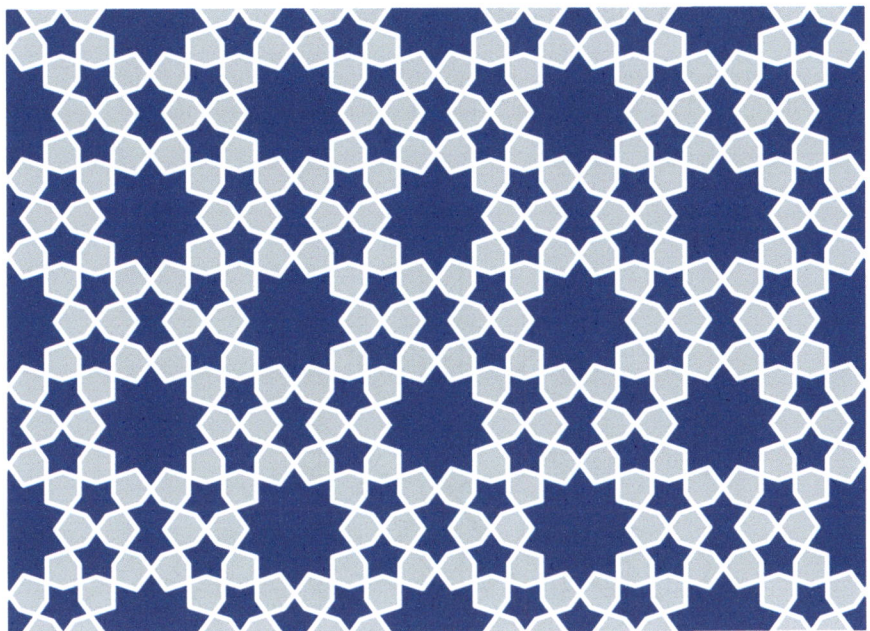

Fig. 4.60 Kok Gumbaz Mosque, Shahrisabz, Uzbekistan

Obtain the final periodic tessellation by multiple translations of the repeat unit (Fig. 4.60).

The tessellation contains convergent standard 10-pointed rosettes whit a central (10,72°) = |10/3| regular star and parallel (5,72°) = |5/1.5| regular stars.

4.1.31 Alhambra Museum, Granada, Spain

One can construct the tessellation of Fig. 4.61 using a rhombic repeat unit. The design of Fig. 4.61 contains 10-pointed parallel ideal rosettes.

Fig. 4.61 Periodic tessellation structure of model 31

4.1 Introduction

Construction of the repeat unit   **Level: Easy**

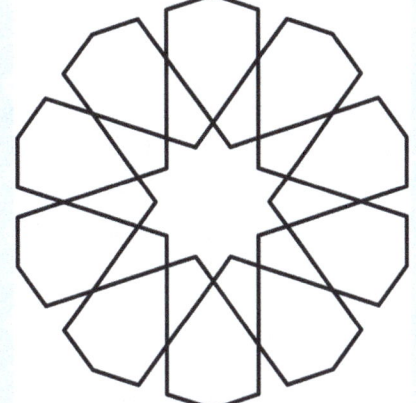

Drawing of the model 31. Step 1
1. Start from a parallel ideal 10-pointed rosette built from a regular star $(10,36°)2 = |10/4|2$.

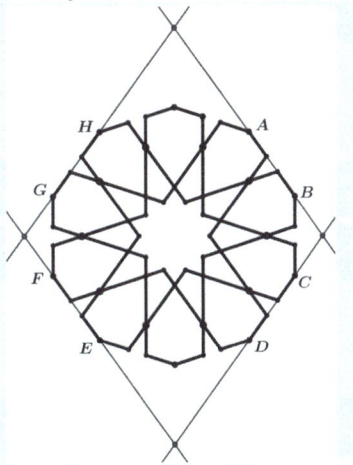

Drawing of the model 31. Step 2
2. The boundary f of the repeat unit is the rhombus defined by the lines passing through the points A and B, C and D, E and F, G and H, respectively.

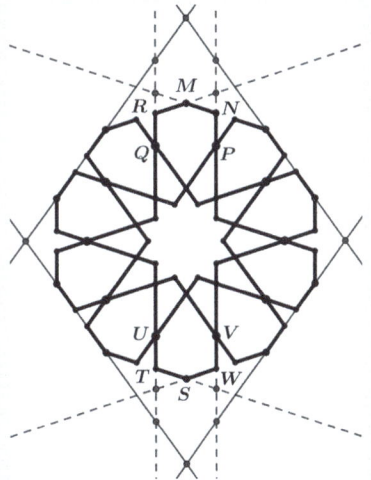

Drawing of the model 31. Step 3
3. Draw the points of intersection of: the ray of origin the point P passing through the point N with the ray of origin the point R passing through the point M and with the boundary f; the ray of origin the point Q passing through the point R with the ray of origin the point N passing through the point M and with f; the ray of origin the point V passing through W with the ray of origin the point T passing through S and with f; the ray of origin the point U passing through the point T with the ray of origin the point W passing through S and with f.

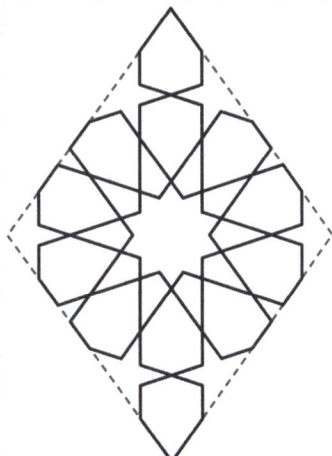

Drawing of the model 31. Step 4
4. Complete the drawing of the part of the tessellation within the interstitial region of the repeat unit.

Fig. 4.62 Alhambra Museum, Granada, Spain

Obtain the final periodic tessellation by multiple translations of the repeat unit (Fig. 4.62).

4.1.32 Shah-i Zinda, Samarkanda, Uzbekistan

One can construct the tessellation of Fig. 4.63 using a repeat unit which is a rectangle.

The design of Fig. 4.63 contains 10-pointed parallel ideal rosettes.

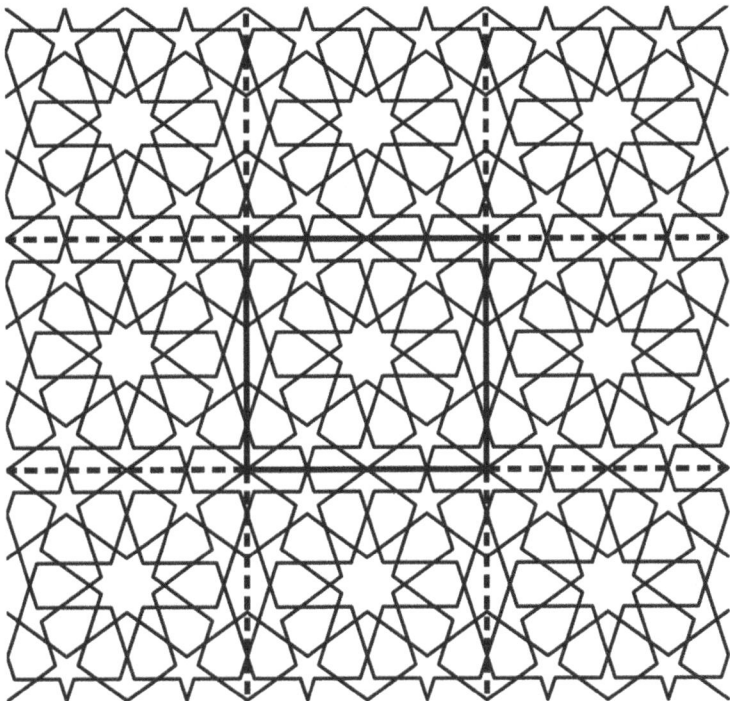

Fig. 4.63 Periodic tessellation structure of model 32

Construction of the repeat unit
Start from a parallel ideal 10-pointed rosette with a central regular star (10,36°)2 = |10/4|2. Draw: the point of intersection R of the line through the points A and B and the line through the points P and Q; the point of intersection S of the line passing through the points G and H and the line passing through the points I and J. The boundary polygon f of the repeat unit is the rectangle of vertices: the point X intersection of the horizontal line passing through R and the line passing through the points C and D; the point Y intersection of the horizontal line passing through the point S and the line passing through the points E and F; the point Z intersection of the horizontal line passing through S and the line passing through the points K and L; the point T intersection of the horizontal line passing through the point R and the line passing through the points M and N.

Level: Medium

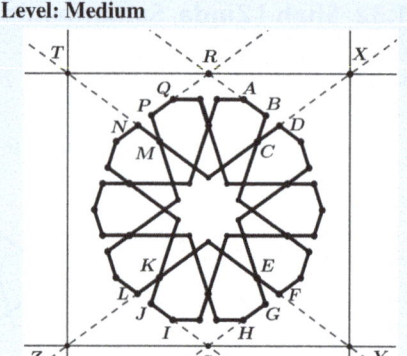

Drawing of the model 32. Step 1

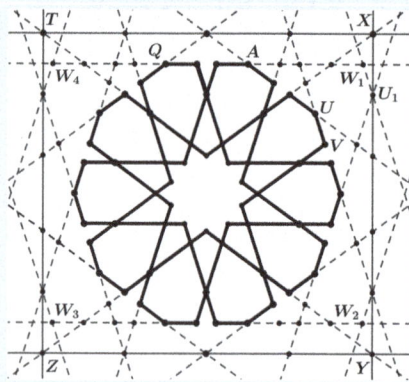

Drawing of the model 32. Step 2
Draw: the extensions toward the boundary polygon f of the sides of the tips of the petals; the points where the extensions intersect and where they intersect f; the point W_1 of the part of the interstitial region close to the vertex X that is the intersection of the line passing through the points A and Q with the line passing through the point U_1 parallel to the line passing through the points U and V. The points W_2, W_3, and W_4 of the parts of the interstitial region close to the vertices Y, Z, and T are drawn in the same way as we did for W_1.

Drawing of the model 32. Step 3
Complete the drawing within the interstitial region of the repeat unit.

4.1 Introduction

Fig. 4.64 Shah-i Zinda, Samarkanda, Uzbekistan

Obtain the final periodic tessellation by multiple translations of the repeat unit (Fig. 4.64).

4.1.33 Amir Azbek al-Yusufi, Cairo, Egypt

One can construct the tessellation of Fig. 4.65 using a repeat unit which is a flattened hexagon.

The design of Fig. 4.65 contains 10-pointed parallel ideal rosettes.

Fig. 4.65 Periodic tessellation structure of model 33

4.1 Introduction

Construction of the repeat unit   **Level: Easy**

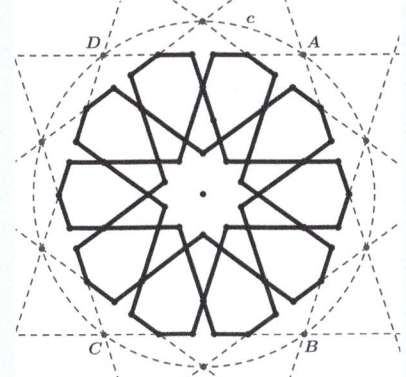

Drawing of the model 33. Step 1
1. Start from an ideal 10-pointed rosette with center a parallel (10.36°)2 = |10/4|2 regular star. Extend the outer sides of the rosette and draw the points A, B, C, and D where they intersect. Draw the circle c passing through A, B, C, and D.

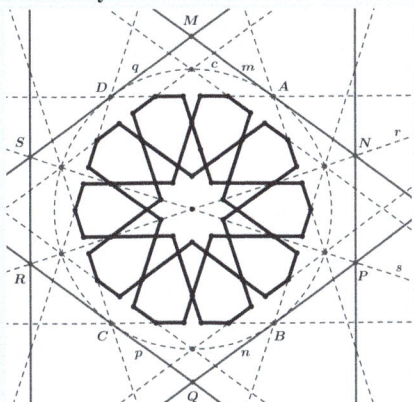

Drawing of the model 33. Step 2
2. Draw the tangents m, n, p, and q to the circle c through the points A, B, C, and D. Draw the points N, P, R, and S intersections of the lines m and r, n and s, p and r, q and s, respectively. The boundary polygon of the repeat unit is the flattened hexagon of vertices M, N, P, Q, R, and S.

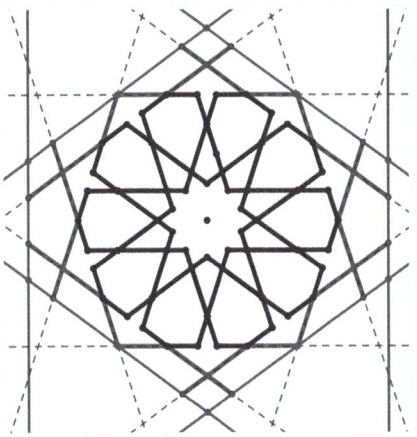

Drawing of the model 33. Step 3
3. Complete the drawing within the interstitial region of the repeat unit.

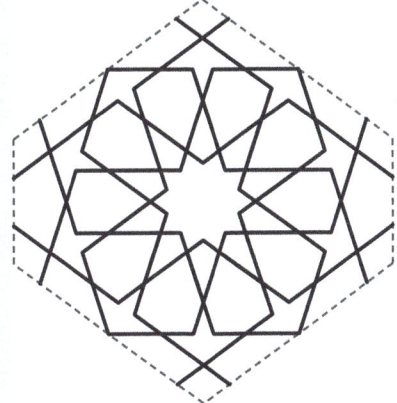

Drawing of the model 33. Step 4
4. Repeat unit

Fig. 4.66 Amir Azbek al-Yusufi, Cairo, Egypt

Obtain the final periodic tessellation by multiple translations of the repeat unit (Fig. 4.66).

4.1.34 Sabil wa Kuttab al-Sultan Qaytbay, Cairo, Egypt

One can construct the tessellation of Fig. 4.67 using a repeat unit which is an elongated hexagon.

The design of Fig. 4.67 contains 10-pointed parallel ideal rosettes.

Fig. 4.67 Periodic tessellation structure of model 34

Construction of the repeat unit — **Level: Medium**

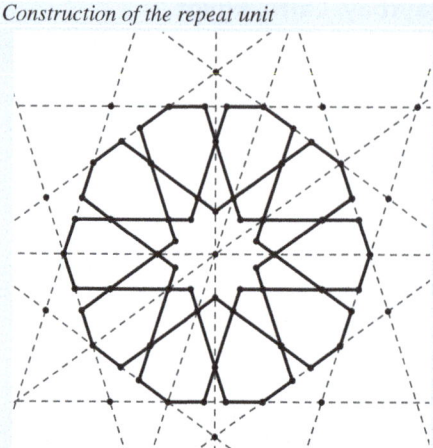

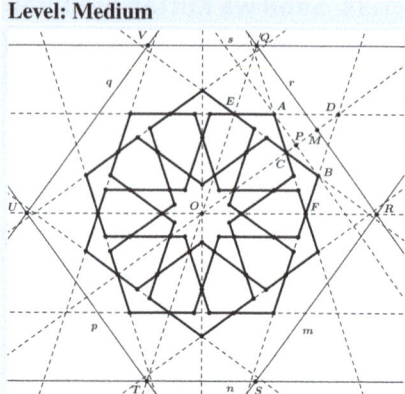

Drawing of the model 34. Step 1
1. Start from a parallel ideal 10-pointed rosette with a central regular star (10,36°)2 = |10/4|2. Draw the lines along the outer sides of the petals of the rosette and their intersections as shown.

Drawing of the model 34. Step 2
2. Draw: the point P intersection of the line passing through the points O and C and the line passing through the points A and B; the point D intersection of the line passing through the points E and A and the line passing through the points F and B; the midpoint M of P and D; the line r passing through M and orthogonal to the line passing through O and C; the point Q intersection of r and the line passing through O and E; the point R intersection of r and the line passing through O and F; the horizontal line s passing through Q; the points S, T, U, and V and the lines m, n, p, and q that define the boundary of the repeat unit using symmetries with respect to the horizontal and vertical lines through O

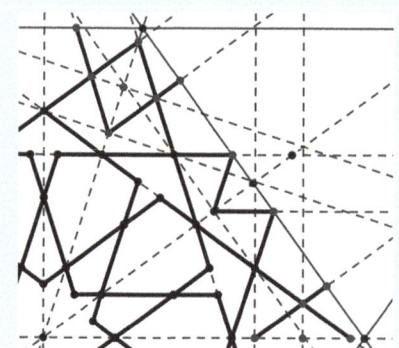

Drawing of the model 34. Step 3
3. Complete the drawing at the top right of the repeat unit as shown.

Drawing of the model 34. Step 4
4. Complete the drawing of the repeat unit using axial symmetries.

4.1 Introduction

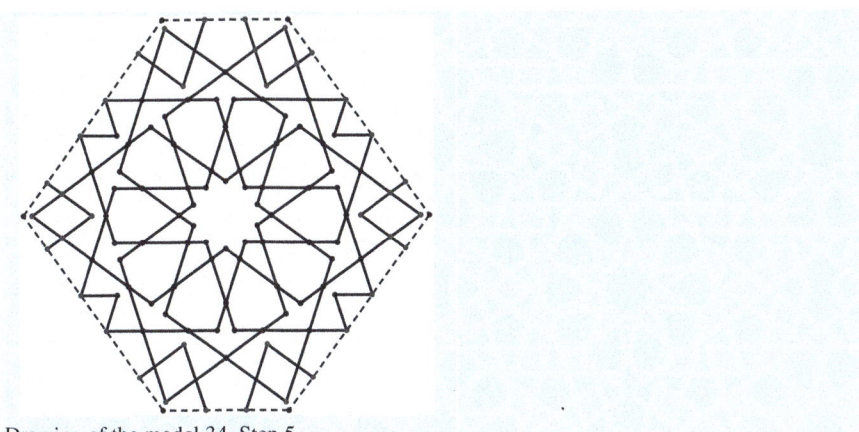

Drawing of the model 34. Step 5
5. The repeat unit.

Fig. 4.68 Sabil wa Kuttab al-Sultan Qaytbay, Cairo, Egypt

Obtain the final periodic tessellation by multiple translations of the repeat unit (Fig. 4.68).

The choice of point P in step 2 determines the size of the rhombuses and their adjacent shapes.

The pattern was historically used in several places with rhombuses of a different size than the one in the model presented here (Wichmann and Wade. 2018).

4.1.35 Ali-Qapu Palace, Isfahan, Iran

One can construct the tessellation of Fig. 4.69 using a repeat unit which is a rectangle.

The design of Fig. 4.69 contains parallel ideal 10-pointed rosettes.

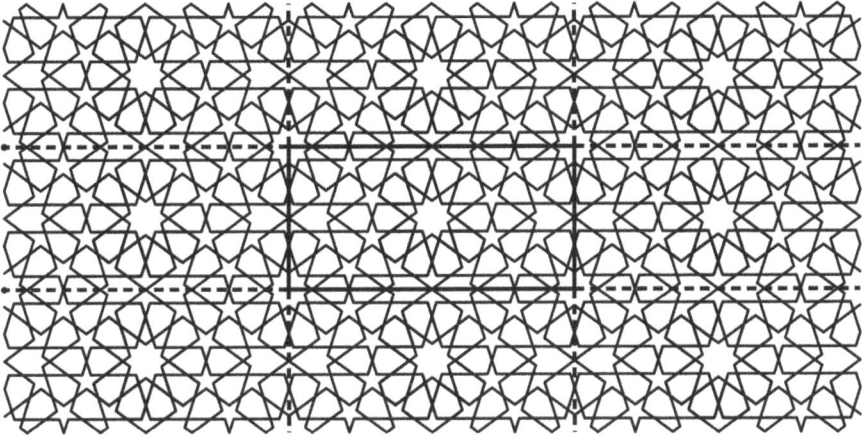

Fig. 4.69 Periodic tessellation structure of model 35

Construction of the repeat unit   **Level: Medium**

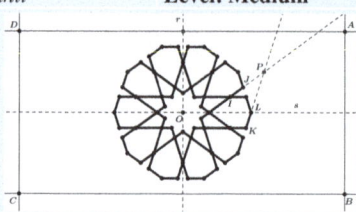

Drawing of the model 35. Step 1
1. Start from a parallel ideal 10-pointed rosette with a central regular star $(10,36°)2 = |10/4|2$. Draw: the point P intersection of the ray of origin I passing through J and the ray of origin K passing through L; the point A symmetrical of the point O with respect to P; the points B, C, and D using symmetries with respect to the horizontal s passing through O and the vertical r passing through O as shown. The points A, B, C, and D are the vertices of the boundary polygon of the repeat unit.

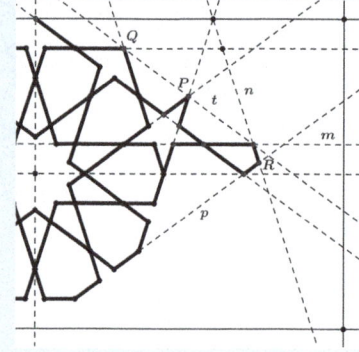

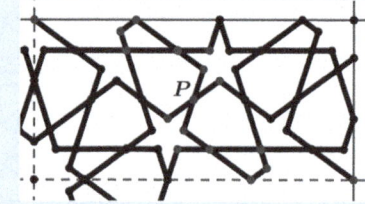

Drawing of the model 35. Step 3
3. Draw the top right of the upper right quadrant of the repeat unit as the symmetrical of the bottom left with respect to the point P.

Drawing of the model 35. Step 2
2. Draw the extensions of the sides of the rosette and the points where they intersect as shown. Draw: the line t passing through the points P and Q; the line n symmetrical of the line m with respect to t; the point R intersection of the lines n and p. Complete the drawing at the bottom left of the top right of the repeating unit as shown.

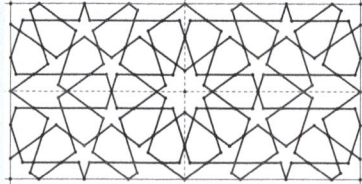

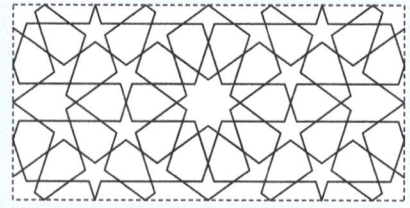

Drawing of the model 35. Step 44.
Complete the drawing of the repeat unit using axial symmetries.

Drawing of the model 35. Step 5
5. The repeat unit.

4.1 Introduction

Fig. 4.70 Ali-Qapu Palace, Isfahan, Iran

Obtain the final periodic tessellation by multiple translations of the repeat unit (Fig. 4.70).

4.1.36 Panel 44 of the Topkapı Scroll, Topkapı Palace museum, Istanbul, Turkey

One can construct the tessellation of Fig. 4.71 using a repeating unit which is a rectangle.

The design of Fig. 4.71 contains parallel ideal 10-pointed and convergent standard 12-pointed rosettes.

Fig. 4.71 Periodic tessellation structure of model 36

4.1 Introduction

Construction of the repeat unit   **Level: Difficult**

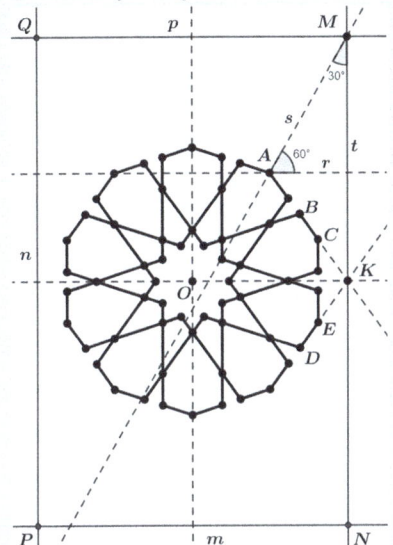

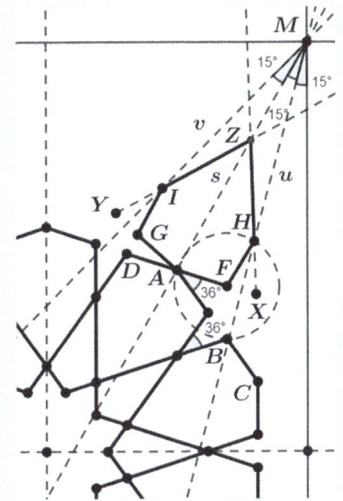

Drawing of the model 36. Step 1

1. Start from a parallel ideal 10-pointed rosette with a central regular star $(10.36°)2 = |10/4|2$. Draw: the horizontal r through the point A; the line s rotated of the line r of center A and 60°; the point K intersection of the ray of origin the point B passing through the point C and the ray of origin the point D passing through E; the vertical line t passing through the point K; the point M intersection of the lines s and t; the horizontal p passing through the point M; the line m symmetric of p with respect to the horizontal passing through the point O; the line n symmetric of t with respect to the vertical passing through O. The points M, N, P, and Q are the vertices of the rectangular repeat unit.

Drawing of the model 36. Step 2

2. First, we will draw a petal of the standard 12-pointed rosette that will be located at the top right of the repeating unit. Draw: the point F symmetric of the point D with respect to the point A; the point G symmetric of F with respect to the line s; the lines u and v rotated of the line s with center the point M and angle of 15° counterclockwise and clockwise; the point X symmetric of F with respect to the line u; the point Y symmetric of G with respect to the line v; the point H intersection of the circle of center F passing through A with the line u; the point I symmetric of H with respect to the line s; the point Z intersection of the ray of origin the point X passing through H and the ray of origin Y passing through I. Draw the petal of vertices A, G, I, Z, H, and F.

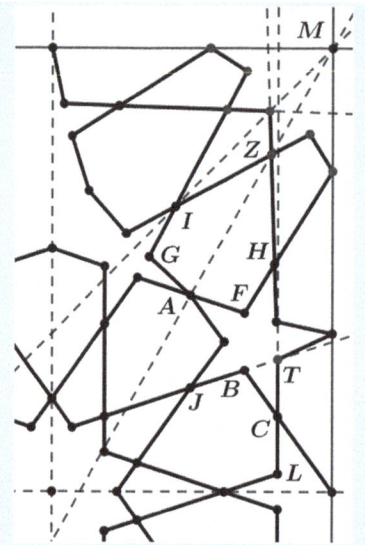

Drawing of the model 36. Step 3
3. Draw the rest of the petals of the 12-pointed rosette located at the top right of the repeating unit rotating the petal AGIZHF with center the point M and angle of 30° counterclockwise and clockwise. Intersect the extension of the petals toward the inside and draw part of the central star. Draw the point T intersection of the ray of origin the point J passing through the point B and the ray of origin the point L passing through C. Draw the missing parts as shown.

Drawing of the model 36. Step 4
4. Complete the repeat unit using axial symmetries.

Drawing of the model 36. Step 5
5. Repeat Unit.

4.1 Introduction

Fig. 4.72 Panel 44 of the Topkapı Scroll, Topkapı Palace museum, Istanbul, Turkey

Obtain the final periodic tessellation by multiple translations of the repeat unit (Fig. 4.72).

The tessellation contains parallel ideal 10-pointed rosettes with a central $(10,36°)2 = |10/4|2$ regular star and convergent standard 12-pointed rosettes determined by the 10-pointed rosettes.

4.1.37 Mosque of Qeycoun, Cairo, Egypt

One can construct the tessellation of Fig. 4.73 using a repeat unit which is a square. The design of Fig. 4.73 contains 12-pointed parallel ideal rosettes.

Fig. 4.73 Periodic tessellation structure of model 37

4.1 Introduction

Construction of the repeat unit

Level: Easy

Drawing of the model 37. Step 1
Start from a 12-pointed parallel ideal rosette built from a regular star (12,30°)2 = |12/5|2.

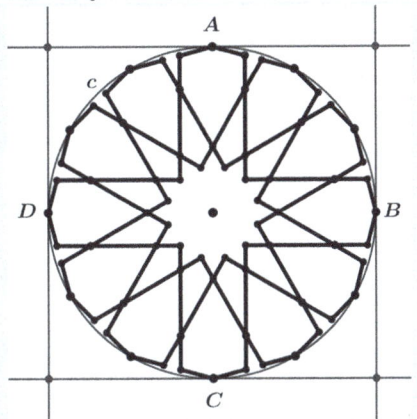

Drawing of the model 37. Step 2
Draw the circle c passing through the tips of the petals of the rosette. The boundary polygon f of the repeat unit is the square bounded by the tangents to the circle c at the tips A, B, C, and D.

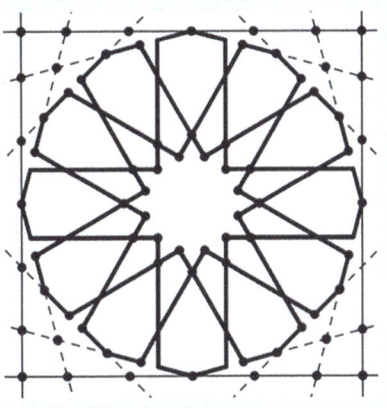

Drawing of the model 37. Step 3
Extend toward the boundary polygon f the sides of the rosette tips that do not belong to f and draw the points where they intersect with f.

Drawing of the model 37. Step 4
Complete the drawing within the interstitial region of the repeat unit.

Fig. 4.74 Mosque of Qeycoun, Cairo, Egypt

Obtain the final periodic tessellation by multiple translations of the repeat unit (Fig. 4.74).

4.1.38 Masjid Suleyman Pasha, Cairo, Egypt

One can construct the tessellation of Fig. 4.75 using a repeat unit which is a regular hexagon.

The design of Fig. 4.75 contains 12-pointed convergent standard rosettes.

Fig. 4.75 Periodic tessellation structure of model 38

Construction of the repeat unit **Level: Easy**

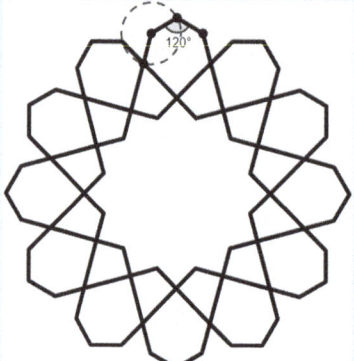

Drawing of the model 38. Step 1
Start from a convergent standard 12-pointed rosette with center a regular star $(12,60°)2 = |12/4|2$. The internal angle of the outer tip of the petals is of 120°, like the internal angles of a regular hexagon.

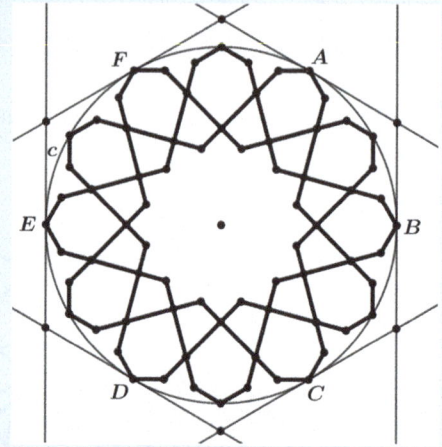

Drawing of the model 38. Step 2
Draw the circle c passing through the tips of the petals of the rosette. The boundary polygon f of the repeat unit is the hexagon bounded by the tangents to the circle c at the tips A, B, C, D, E, and F of the petals of the rosette.

Drawing of the model 38. Step 3
Extend toward the boundary polygon f the sides of the tips of the petals that do not belong to f and draw the points where they intersect with f.

Drawing of the model 38. Step 4
Complete the drawing within the interstitial region of the repeat unit.

4.1 Introduction

Fig. 4.76 Masjid Suleyman Pasha, Cairo, Egypt

Obtain the final periodic tessellation by multiple translations of the repeat unit (Fig. 4.76).

4.1.39 Mihrab of the Great Mosque, Damascus, Syria

One can construct the tessellation of Fig. 4.77 using a repeat unit which is a regular hexagon.

The design of Fig. 4.77 contains 12-pointed parallel ideal rosettes.

Fig. 4.77 Periodic tessellation structure of model 39

4.1 Introduction

Construction of the repeat unit

Level: Easy

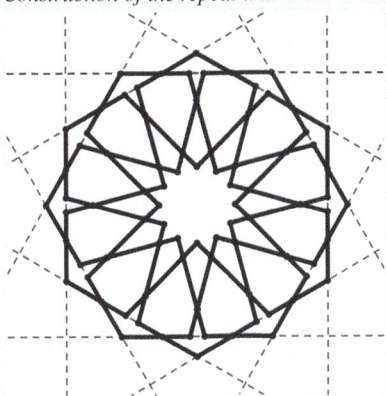

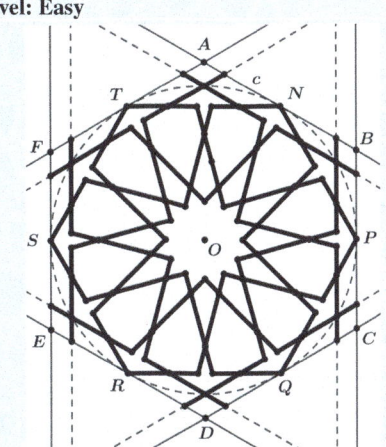

Drawing of the model 39. Step 1
Start from a parallel ideal 12-pointed rosette with center a regular star (12,30°)2 = |12/5|2. Draw the extension of the sides of the petals to where they intersect as shown.

Drawing of the model 39. Step 2
Draw: the circle c of center the point O passing through the points N, P, Q, R, S, and T. Draw the tangents to the circle c at the points N, P, Q, R, S, and T. The intersection points A, B, C, D, E, and F of the tangents are the vertices of the boundary polygon of the repeat unit. Complete the design inside the interstitial region as shown.

Drawing of the model 39. Step 3
Repeat Unit.

Fig. 4.78 Mihrab of the Great Mosque, Damascus, Syria

Obtain the final periodic tessellation by multiple translations of the repeat unit (Fig. 4.78).

4.1.40 Marbel Panel, Great Mosque, Damascus, Syria

One can construct the tessellation of Fig. 4.79 using a repeat unit which is a regular hexagon.

The design of Fig. 4.79 contains regular convergent 12-pointed rosettes and divergent regular 4-pointed rosettes.

Fig. 4.79 Periodic tessellation structure of model 40

Construction of the repeat unit
1. Start from a convergent regular 12-pointed rosette with a central $(12,60°)2 = |12/4|2$ regular star, and an internal angle of the outer tip of the petals of 150°. Draw: the point S intersection of the ray of origin the point P passing through the point Q with the circle of center Q passing through the point R; the point T symmetric of the point S with respect to the line passing through the points O and R; the line m passing through the points S and T; the lines n, p, q, r, and s obtained from the line m with successive rotations of center O an angle of 60°. The points A, B, C, D, E, and F intersection of the lines s and m, m and n, n and p, p and q, q and r, r and s are the vertices of the boundary polygon f of the repeat unit.

Level: Easy

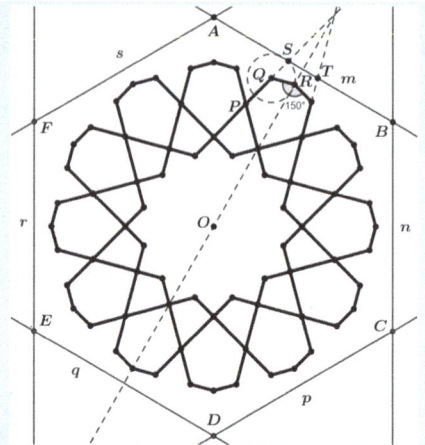

Drawing of the model 40. Step 1

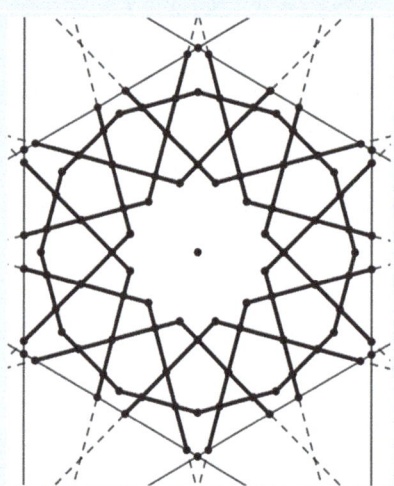

Drawing of the model 40. Step 2

2. Draw the extension of the sides of the petals and the points where they intersect the boundary polygon f. Complete the drawing within the interstitial region to obtain the repeating unit as shown.

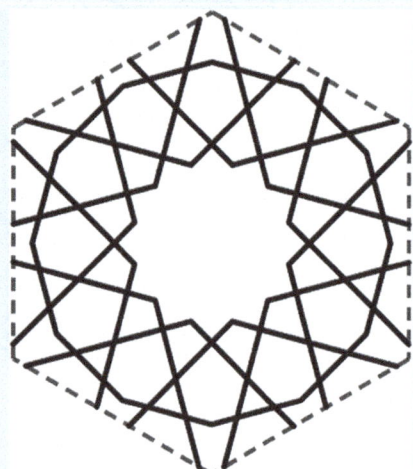

Drawing of the model 40. Step 3

3. The repeat unit.

4.1 Introduction

Fig. 4.80 Marbel panel, Great Mosque, Damascus, Syria

Obtain the final periodic tessellation by multiple translations of the repeat unit (Fig. 4.80).

The resulting tessellation contains convergent regular 12-pointed rosettes with a central $(12.60°)2 = |12/4|2$ regular star and divergent regular 4-pointed rosettes with a central $(4.60°) = |4/(4/3)|$ regular star.

4.1.41 Al-Maridani Mosque, Cairo, Egypt

One can construct the tessellation of Fig. 4.81 using a repeat unit which is a square.

The design of Fig. 4.81 contains regular convergent 12-pointed rosettes and divergent regular 4-pointed rosettes.

Fig. 4.81 Periodic tessellation structure of model 41

4.1 Introduction

Construction of the repeat unit **Level: Medium**

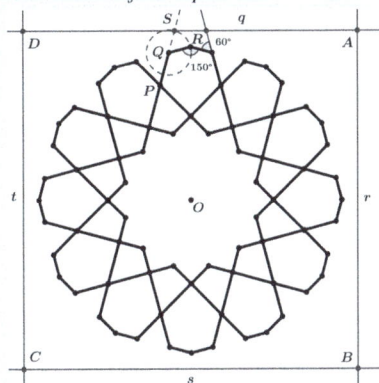

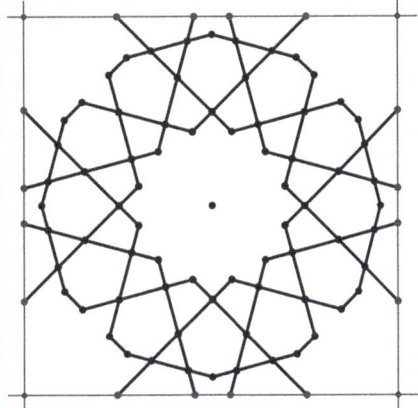

Drawing of the model 41. Step 1
1. Start from a convergent regular 12-pointed rosette with a central (12,60°)2 = |12/4|2 regular star, and an internal angle of the outer tip of the petals of 150°. Draw: the point S intersection of the ray of origin the point P passing through the point Q with the circle of center Q passing through the point R; the horizontal q passing through S; the lines r, s, and t obtained from q with successive rotations of center the point O an angle of 90°; the points A, B, C, and D intersection of the lines q and r, r and s, s and t, t and q, that are the vertices of the boundary polygon f of the repeat unit.

Drawing of the model 41. Step 2
2. Draw the extension of some of the sides of the petals as shown. Draw the points where the extensions intersect, which are the vertices of the boundary polygon f. Draw parts of the model as shown.

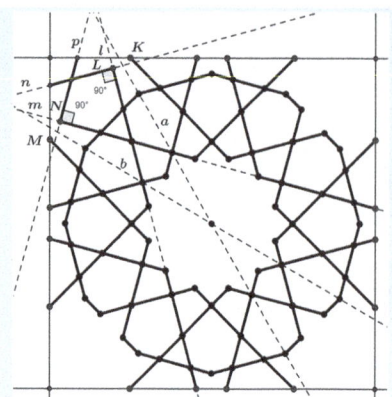

Drawing of the model 41. Step 3
3. Draw: the point L symmetric of the point K with respect to the line a; the point N symmetric of the point M with respect to the line b; the line n rotated of the line l with a rotation of center L and angle 90°; the line p rotated of the line m with a rotation of center the point N and angle 90°. Draw part of the model as shown.

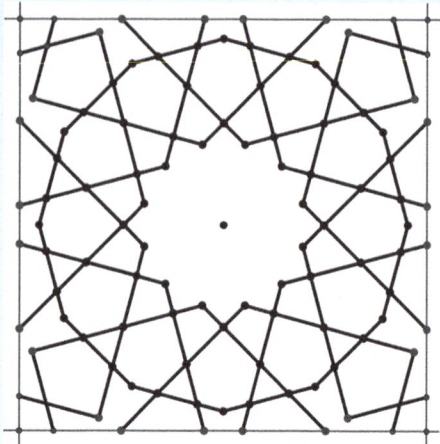

Drawing of the model 41. Step 4
4. Complete the drawing within the interstitial region to obtain the repeat unit.

4.1 Introduction

Drawing of the model 41. Step 5
5. The repeat unit

Fig. 4.82 Al-Maridani Mosque, Cairo, Egypt

Obtain the final periodic tessellation by multiple translations of the repeat unit (Fig. 4.82).

The resulting tessellation contains convergent regular 12-pointed rosettes with a central (12.60°)2 = |12/4|2 regular star and divergent regular 4-pointed rosettes with a central (4.60°) = |4/(4/3)| regular star.

4.1 Introduction

4.1.42 Gur-e-Amir Mausoleum, Samarkand, Uzbekistan

One can construct the tessellation of Fig. 4.83 using a repeat unit which is a square.

The design of Fig. 4.83 contains convergent standard 12-pointed rosettes and parallel ideal 8-pointed rosettes.

Fig. 4.83 Periodic tessellation structure of model 42

Construction of the repeat unit **Level: Difficult**

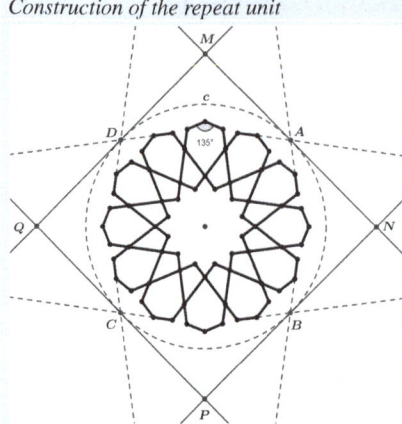
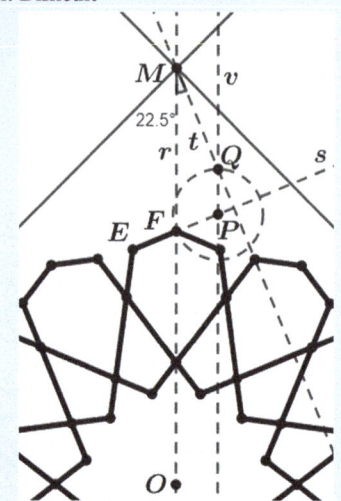

Drawing of the model 42. Step 1
1. Start from a convergent standard 12-pointed rosette with a central I(12,45°)2 = I12,4.5I2 regular star, and an internal angle of the outer tip of the petals of 135°. Draw: the extension of the outer sides of the rosette as shown; the points A, B, C, and D where the extensions intersect; the circle c passing through the points A, B, C, and D; the tangents to the circle c passing through the points A, B, C, and D; the intersection points M, N, P, and Q of the tangents. The boundary polygon of the repeat unit is the square of vertices M, N, P, and Q.

Drawing of the model 42. Step 2
2. In the upper part of the interstitial region, we must draw part of an ideal rosette based in a parallel star I8.3I2 = (8.45°)2. Draw: the vertical r passing through the point O; the line t rotated of r an angle of 22.5° counterclockwise around the point M. Draw the point P on the ray s of origin the point E passing through the point F so that: the line t and the vertical v passing through P intersect at a point Q of the circle with center P passing through F.

4.1 Introduction

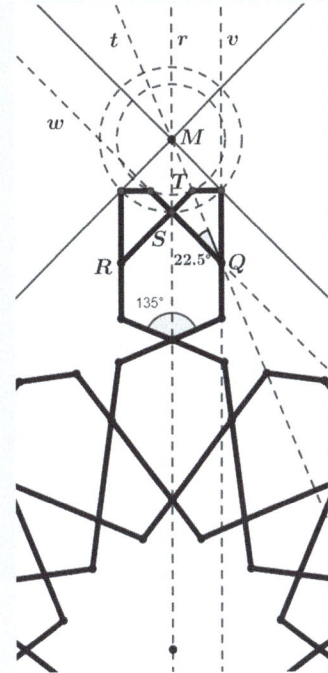

Drawing of the model 42. Step 3
3. Draw: the point R symmetric of the point Q with respect to the line r; the line w rotated of the line t around Q an angle of 22.5° counterclockwise; the point S intersection of lines r and w; the point T intersection of the line passing through R and S with t. Complete the drawing of the part of the 8-point rosette as shown.

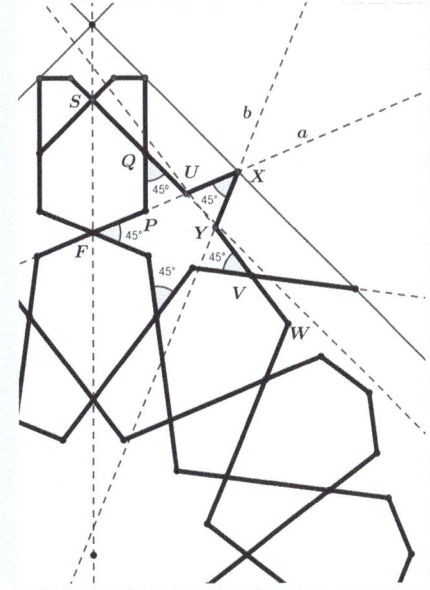

Drawing of the model 42. Step 4
4. Draw: the line a through F and P; the point U intersection of the ray of origin S through Q and the line a; the point X intersection of line a with the side of the boundary polygon; the line b rotated of the line a around the point X an angle of 45° counterclockwise; point Y intersection of the line b with the ray of origin the point W passing through the point V. Complete the drawing in the upper right interstitial region as shown.

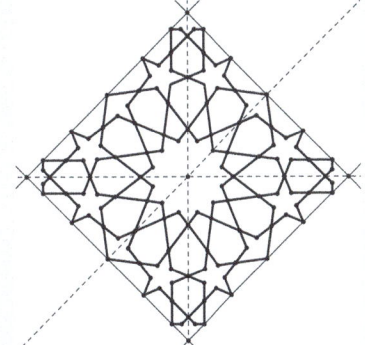

Drawing of the model 42. Step 5
5. Complete the drawing of the repeat unit with successive axial symmetries.

Drawing of the model 42. Step 6
6. Repeat unit.

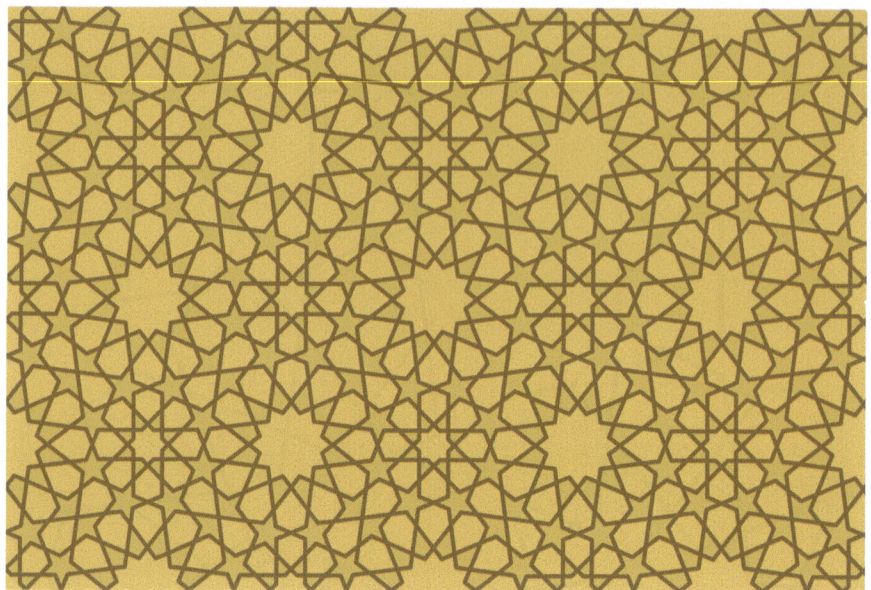

Fig. 4.84 Gur-e-Amir Mausoleum, Samarkand, Uzbekistan

Obtain the final periodic tessellation by multiple translations of the repeat unit (Fig. 4.84).

The tessellation contains convergent standard 12-pointed rosettes of center a (12,45°)2 = | 12/4.5|2 regular star, parallel ideal 8-pointed rosettes of center a (8,45°)2 = |8/3|2 regular star, and (5,45°) divergent regular stars.

4.1 Introduction

4.1.43 Attarine Medersa, Fez, Morocco

One can construct the tessellation of Fig. 4.85 using a repeat unit which is a regular hexagon.

The design of Fig. 4.85 contains parallel 12-pointed rosettes and parallel 8-pointed stars.

Fig. 4.85 Periodic tessellation structure of model 43

Construction of the repeat unit **Level: Difficult**

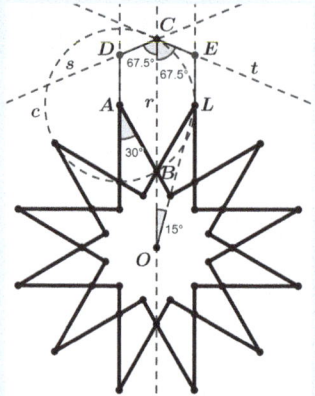
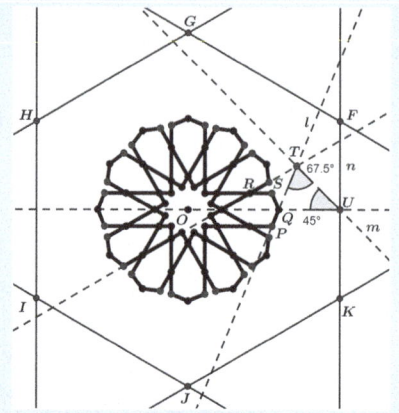

Drawing of the model 43. Step 1
1. Start from a $(12,30°)2 = |12/5|2$ regular star. Draw: the intersection C of the vertical r passing through O with the circle of center A passing through B; the intersection D of the vertical passing through A with the line s rotated of r by a rotation of center C and 67.5° clockwise; the intersection E of the vertical passing through L with the line t rotated of r by a rotation of center C and 67.5° counterclockwise; the point L. Points B, L, E, C, D, and A are the vertices of the upper petal of the rosette.

Drawing of the model 43. Step 2
2. Complete the drawing of the parallel 12-pointed rosette with successive rotations of the upper petal with center O and angle of 60°. Draw: the point T intersection of the line l passing through P and Q and the line passing through R and S; the line m rotated of l by a rotation of center T and angle of 67.5°; the point U intersection of m with the horizontal passing through O; the vertical n passing through U; the boundary polygon of vertices F, G, H, I, J, and K by successive rotations of n of center O and angle of 60°.

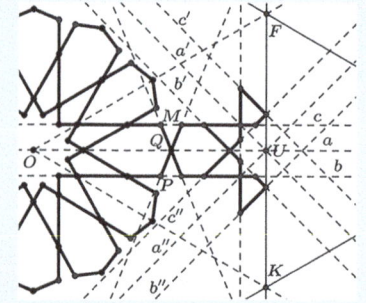
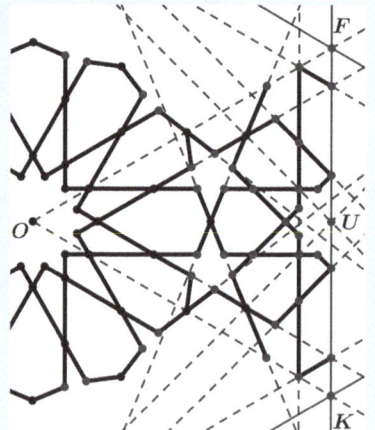

Drawing of the model 43. Step 3
3. Draw: the horizontal lines a, b, and c passing through the points Q, P, and M; the lines a′, b′, and c′ rotated of a, b, and c with a rotation of center U and angle of 45° clockwise; the lines a″, b″, and c″ rotated of a, b, and c with a rotation of center U and angle of 45° counterclockwise; Complete the drawing of the half 8-pointed star and its petal as shown.

Drawing of the model 43. Step 4
4. Extend some sides of the 12-pointed rosette and of the 8-pointed star and its petal and complete the drawing in the triangular right part of the repeat unit as shown.

4.1 Introduction

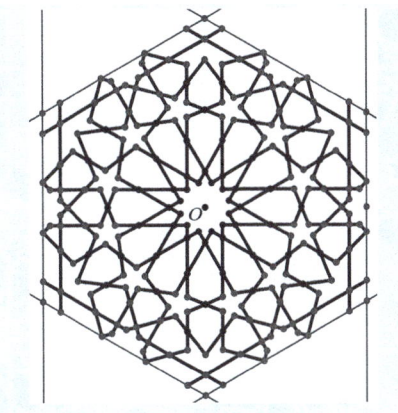

Drawing of the model 43. Step 5
5. Complete the drawing within the interstitial region by successive rotations of center O and angle of 60° of the triangular right part of the repeat unit drawn in steps 3 and 4.

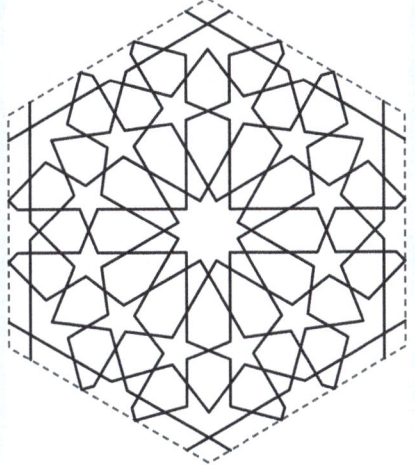

Drawing of the model 43. Step 6
6. The repeat unit.

Fig. 4.86 Attarine Medersa, Fez, Morocco

Obtain the final periodic tessellation by multiple translations of the repeat unit (Fig. 4.86).

The resulting tessellation contains parallel 12-point rosettes with a central $(12,30°)2 = |12/5|2$ regular star and parallel $(8,45°)2 = |8/3|2$ regular stars.

4.1 Introduction

4.1.44 Al-Muayyad Mosque, Cairo, Egypt

One can construct the tessellation of Fig. 4.87 using a repeat unit which is a regular hexagon.

The design of Fig. 4.87 contains standard 12-pointed rosettes.

Fig. 4.87 Periodic tessellation structure of model 44

Construction of the repeat unit **Level: Medium**

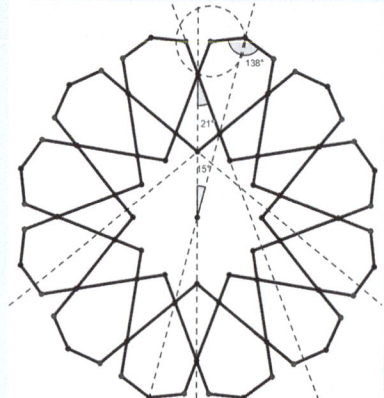
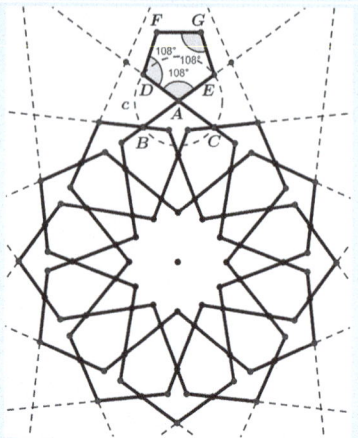

Drawing of the model 44. Step 1
1. Start from a standard 12-pointed rosette with center a regular star $(12,42°)2 = |12/4.6|2$ and internal angle of the outer tip of the petals of 138°.

Drawing of the model 44. Step 2
2. Draw: the lines along the outer sides of the petals of the rosette, their points of intersection, and the segments joining them as shown; the circle c of center A through points B and C; the points D and E intersection of c with the rays of origin C through A and of origin B through A, respectively; the point F rotation of A with center D and 108° counterclockwise; the point G rotation of A with center E and 108° clockwise; the regular pentagon of vertices A, E, G, F, and D.

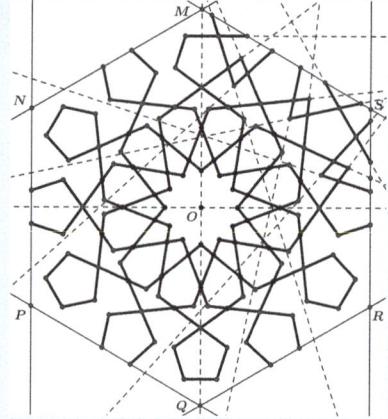

Drawing of the model 44. Step 3
3. Draw the missing pentagons, some without a side as shown, with successive rotations of center O and 30° of the pentagon drawn in step 2.

Drawing of the model 44. Step 4
4. Draw the regular hexagon, boundary of the repeat unit, of vertices M, N, P, Q, R, and S as shown. Complete the drawing of the interstitial region in the upper right of the repeating unit by extending the sides of the pentagons as shown.

4.1 Introduction

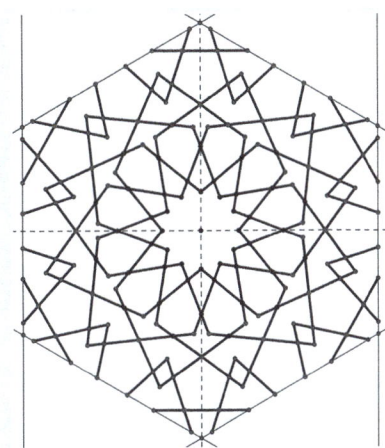

Drawing of the model 44. Step 5
5. Complete the drawing within the interstitial region of the repeat unit using axial symmetries.

Drawing of the model 44. Step 6
6. Repeat unit.

Fig. 4.88 Al-Muayyad Mosque, Cairo, Egypt

Obtain the final periodic tessellation by multiple translations of the repeat unit (Fig. 4.88).

4.1.45 Abd al-Ghani al-Fakhri mosque, Cairo, Egypt

One can construct the tessellation of Fig. 4.89 using a repeat unit which is a flattened hexagon.

The design of Fig. 4.89 contains parallel ideal 14-pointed rosettes and parts of convergent ideal 7-pointed rosettes.

Fig. 4.89 Periodic tessellation structure of model 45

Construction of the repeat unit **Level: Difficult**

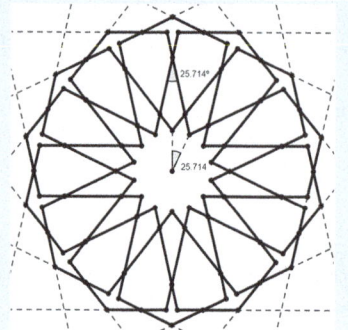

Drawing of the model 45. Step 1
1. Start from a parallel ideal 14-pointed rosette of center a $(14, 25.714°)2 = |14/6|2$ regular star. Draw: the lines along the outer sides of the petals of the rosette, their points of intersection, and the segments joining them as shown. We have approximated 360°/14 by 25.714°.

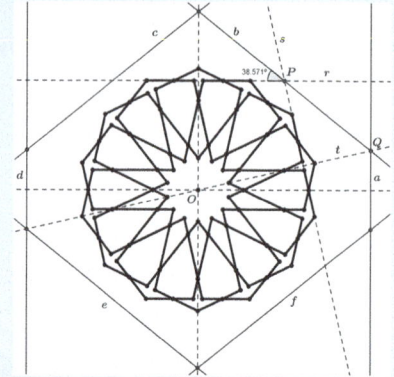

Drawing of the model 45. Step 2
2. Draw: point P intersection of lines r and s; line b rotated of r of center P and 38.571°; point Q intersection of lines b and t; the vertical a through Q; lines c, d, e, and f by using symmetries with respect to the horizontal and vertical lines through O. Lines a, b, c, d, e, and f determine the boundary polygon of the repeat unit. We have approximated 1.5*360°/7 by 38.571°.

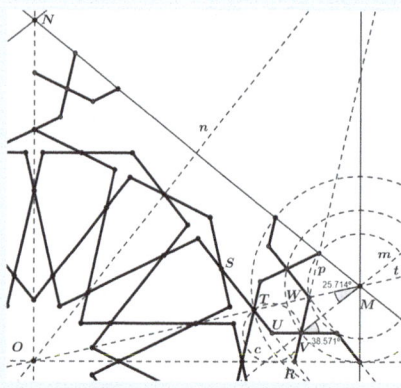

Drawing of the model 45. Step 3
3. Draw: point R intersection of the ray of origin point S passing through T with the horizontal through O; line m rotated of t of center M and 25.714° counterclockwise; point U symmetrical of R with respect to m; circle c of center U passing through T; point V intersection of m and c; ray p of origin R through V; point W intersection of p and t. Complete part of the 7-pointed rosette of center M as shown. Draw part of the 7-pointed rosette of center N by a symmetry with respect to line n of the top half of the part of the rosette of center M.

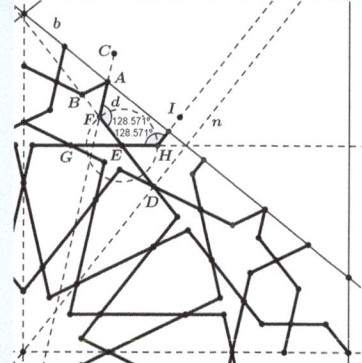

Drawing of the model 45. Step 44. We want to draw half of a regular heptagon at the top right of the repeat unit. Draw: point C symmetrical of point B with respect to the line b; point F intersection of the ray of origin C through A and the ray of origin D through E; the circle d of center E passing through F; point H intersection of d with the ray of origin G passing through E; point I symmetrical of point H with respect to the line b. Complete the drawing of the half heptagon as shown. Draw another half heptagon by a symmetry with respect to the line n of the first half heptagon drawn.

4.1 Introduction

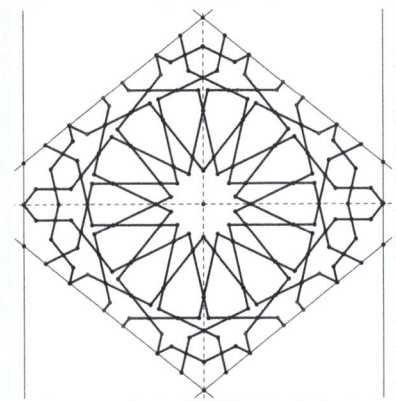

Drawing of the model 45. Step 5
5. Complete the drawing of the repeat unit using axial symmetries.

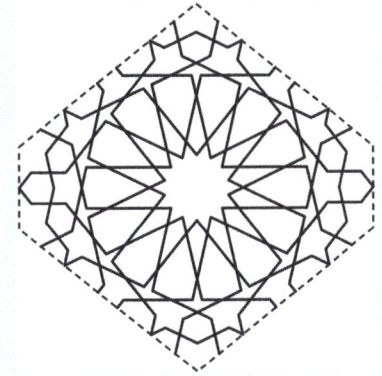

Drawing of the model 45. Step 6
6. The repeat unit.

Fig. 4.90 Abd al-Ghani al-Fakhri mosque, Cairo, Egypt

Obtain the final periodic tessellation by multiple translations of the repeat unit (Fig. 4.90).

The tessellation contains parallel ideal 14-pointed rosettes of center a |14/6|2 = (14,25.714°)2 regular star and parts of convergent ideal 7-pointed rosettes of center a |7/2| = (7,77.143°) regular star.

4.1.46 Sultan Barsbay Funerary Complex, Cairo, Egypt

One can construct the tessellation of Fig. 4.91 using a repeat unit which is a square. The design of Fig. 4.91 contains parallel ideal 8-pointed and 16-pointed rosettes.

Fig. 4.91 Periodic tessellation structure of model 46

Construction of the repeat **Level: Difficult**

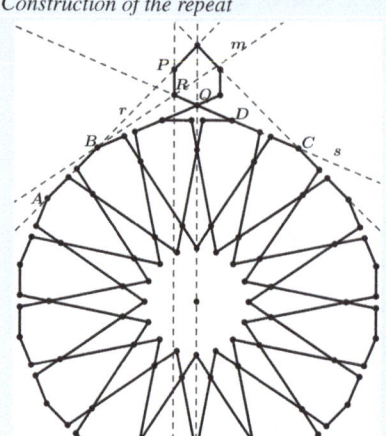
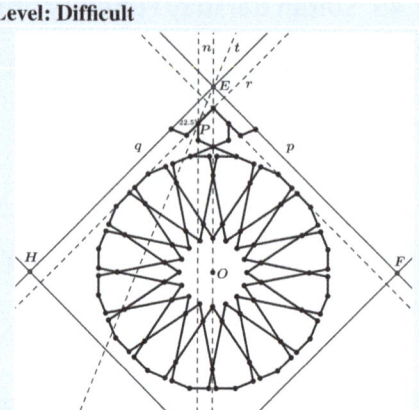

Drawing of the model 46. Step 1
1. Start from a parallel ideal 16-pointed rosette with a central (16,22.5°)2 = |16/7|2 regular star. At the top of the rosette we will draw the petal of an parallel ideal 8-pointed rosette of center a (8.45°)2 = | 8.3|2 regular star. Draw the point P on the line r passing through the points A and B so that the vertical through P and the line s through the points C and D intersect at a point R of the perpendicular bisector m of P and Q.

Drawing of the model 46. Step 2
2. Draw: the line t rotation of the line r of center the point P and angle 22.5° counterclockwise; the vertex E of the boundary polygon intersection of the line t and the vertical n passing through the point O; the two half-petals rotating the vertices of the known petal with rotations of center the point E and 45° clockwise and counterclockwise. Complete the boundary polygon of vertices E, F, G, and H as shown.

4.1 Introduction

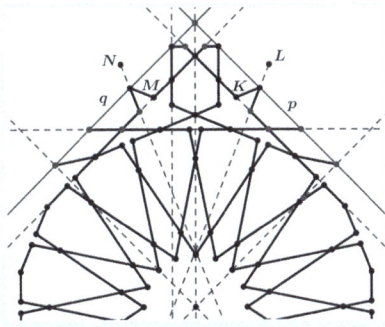

Drawing of the model 46. Step 3
3. Draw: the points L and N symmetric of the points K and M with respect to the lines p and q; the ray of origin L through K; the ray of origin N through M; the extension of the sides of the petals. Complete the drawing of the part of the model corresponding to the top of the interstitial region as shown.

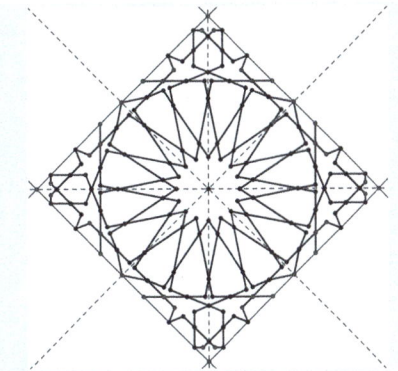

Drawing of the model 46. Step 4
4. Complete the drawing of the repeat unit by using axial symmetries.

Drawing of the model 46. Step 5
5. Repeat unit.

Fig. 4.92 Sultan Barsbay funerary complex, Cairo, Egypt

Obtain the final periodic tessellation by multiple translations of the repeat unit (Fig. 4.92).

The resulting tessellation contains parallel ideal 16-pointed rosettes with a central $(16,22.5°)2 = |16/7|2$ regular star and parallel ideal 8-pointed rosettes of center a $(8,45°)2 = |8/3|2$ regular star.

4.1.47 Niche in the Comares Palace, Alhambra, Granada, Spain

One can construct the tessellation of Fig. 4.93 using a repeat unit which is a square.

The design of Fig. 4.93 contains standard divergent 4-pointed rosettes and 8-pointed regular stars.

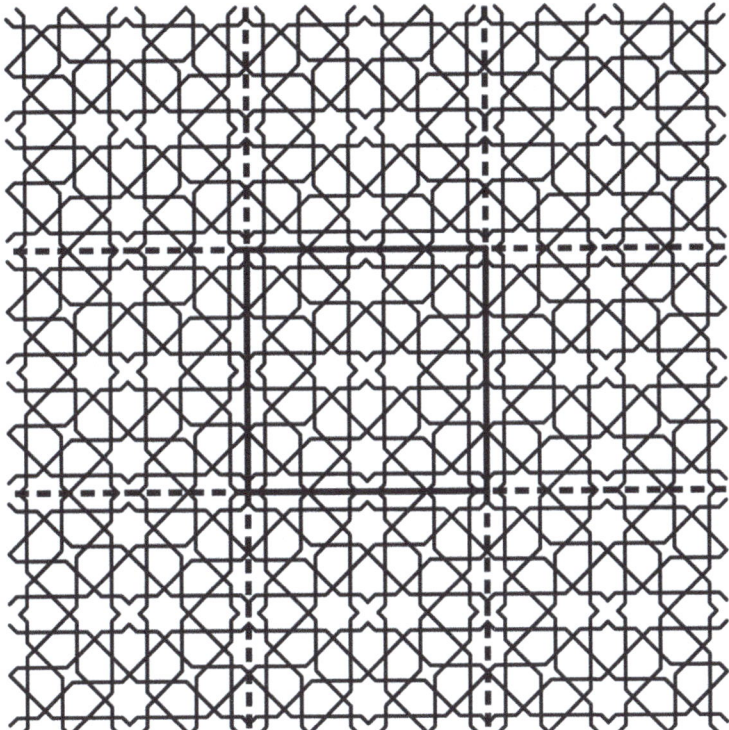

Fig. 4.93 Periodic tessellation structure of model 47

Construction of the repeat unit

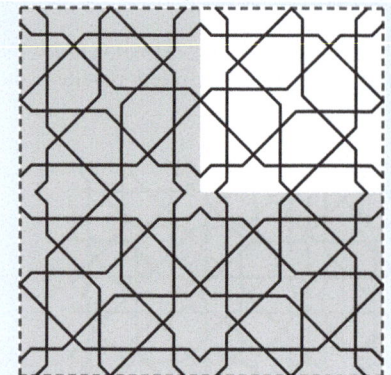

Base unit and repeat unit
We will begin by drawing the square base unit located in the upper right part of the repeat unit, and then complete it using axial symmetries.

Level: Easy

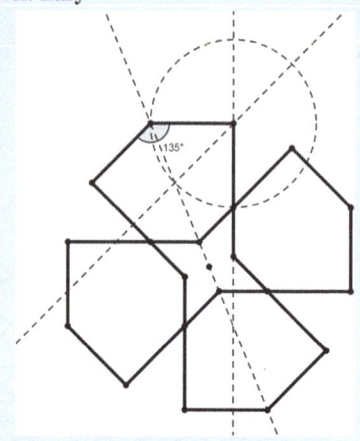

Drawing of the model 47. Step 1
1. Start from a regular star $(4,45°) = |4/1.5|$. Draw a divergent standard 4-pointed rosette with the angle of the outer vertex of the petals of $135°$.

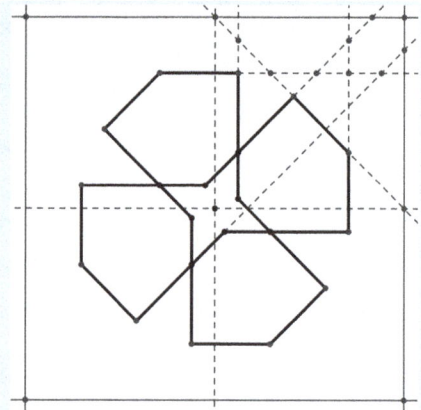

Drawing of the model 47. Step 2
2. Draw the boundary polygon of the base unit of vertices A, B, C, and D as shown.

Drawing of the model 47. Step 3
3. Extend the sides of the petals in the upper right square of the interstitial region as shown.

4.1 Introduction

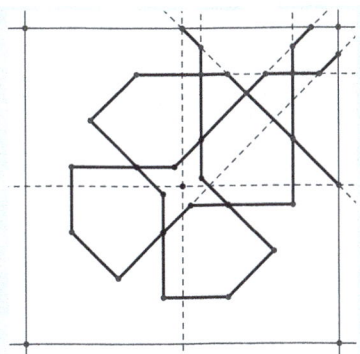

Drawing of the model 47. Step 4
4. Draw the parts of the design in the upper right square of the interstitial region.

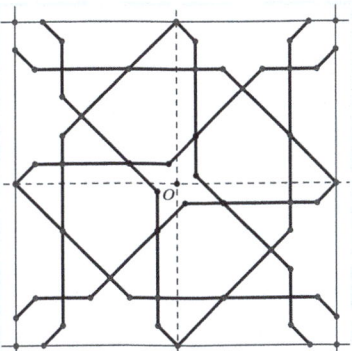

Drawing of the model 47. Step 5
5. Complete the drawing within the interstitial region of the base unit by successive rotations of center O an angle of 90°.

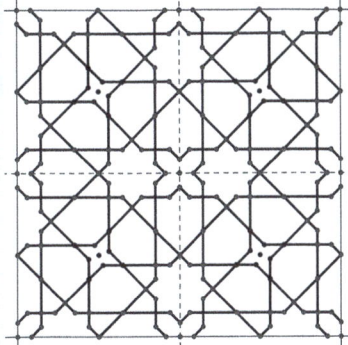

Drawing of the model 47. Step 6
6. Finish drawing the repeat unit using axial symmetries.

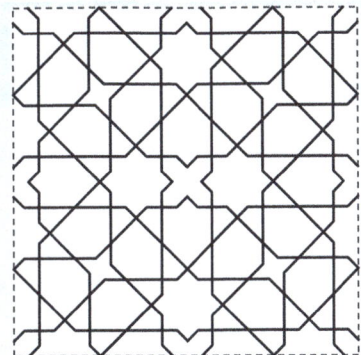

Drawing of the model 47. Step 7
7. The repeating unit in addition to the standard divergent 4-pointed rosettes contains regular stars| 8,2| = (8.90°).

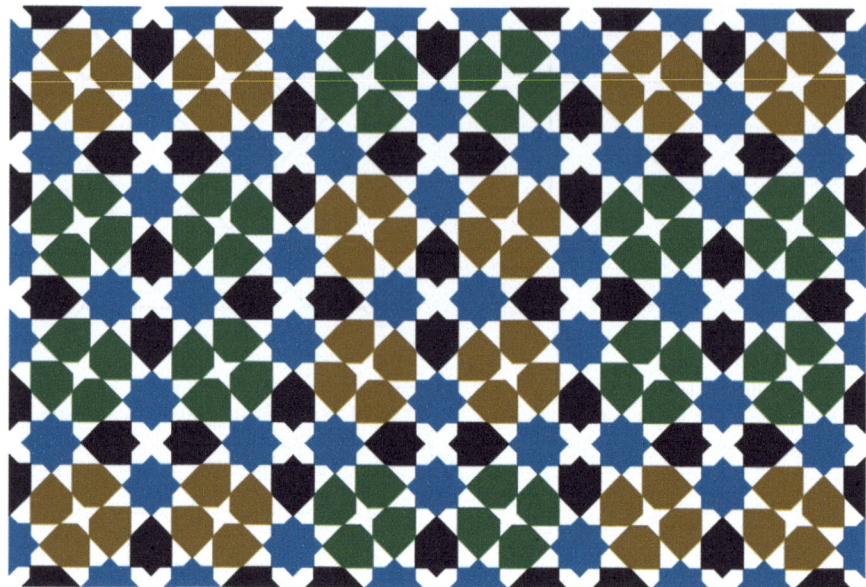

Fig. 4.94 Niche in the Comares Palace, Alhambra, Granada, Spain

Obtain the final periodic tessellation by multiple translations of the repeat unit (Fig. 4.94).

The tessellation contains divergent standard 4-pointed rosettes with center a regular star (4,45°) = |4/1.5| and (8,90°) = |8/2| regular stars.

4.1.48 Two Sisters Room, Alhambra, Granada, Spain

One can construct the tessellation of Fig. 4.95 using a repeat unit which is a square.

The design of Fig. 4.95 contains parallel 16-pointed rosettes, ideal parallel 8-pointed rosettes, and convergent 8-pointed regular stars.

Fig. 4.95 Periodic tessellation structure of model 48

Construction of the repeat unit

Base unit and repeat unit
We will start by drawing the triangular unit base located at the left upper half of the upper right quadrant of the square repeat unit, and then complete it using axial symmetries.

The width of the petals of the 8-pointed rosette is the same as the width of the petals of the 16-pointed rosette and also the same of the figures just outside the 8-pointed rosette included the 8-pointed regular stars. The sides of the 8-pointed rosette and the 8-pointed regular star extend to the petals of the 16-pointed rosette. Thus, the 16-pointed rosette is highly constrained.

Level: Difficult

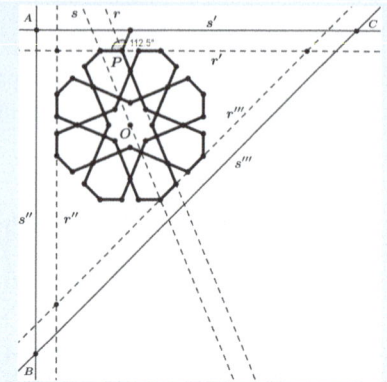

Drawing of the model 48. Step 1
1. Start from a parallel ideal rosette based on a $(8, 45°)2 = |8/3|2$ regular star. Draw lines: r' rotation of line r of center the point P and 112.5° clockwise; s' rotation of line s of center P and 112.5° clockwise; r'' rotation of r' of center the point O and 90° counterclockwise; s'' rotation of s' of center O and 90° counterclockwise; r''' rotation of r'' of center O and 135° counterclockwise; s''' rotation of s'' of center O and 135° counterclockwise. The base unit boundary f is the right triangle of vertices A intersection of s' and s'', B of s'' and s''', and C of s''' and s'.

4.1 Introduction

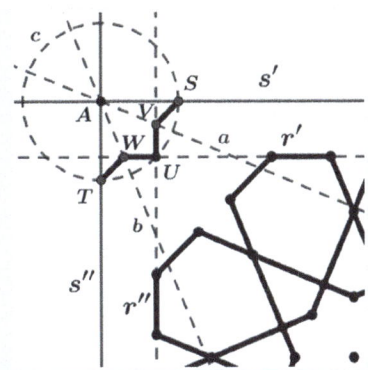

Drawing of the model 48. Step 2

2. The part of the |8/2| = (8.90°) regular star of the upper left side of the repeat unit is determined by the lines r', s', r", and s". Draw: the point A intersection of the lines s' and s"; the point U intersection of the lines r' and r"; the circle c with center A passing through U; the points S and T intersection of c with s' and s"; the perpendicular bisectors a and b of the segments of endpoints S and U, and U and T; the points V and W intersection of lines a and b with lines r' and r". The points S, V, U, W, and T are the vertices of the part of the |8/2| = (8,90°) regular star of the upper left side of the repeat unit.

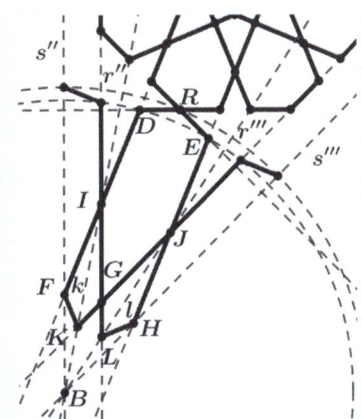

Drawing of the model 48. Step 3

3. Draw: the points D and E intersecting the extension of the sides of the petal of outer vertex R of the 8-pointed rosette; the missing vertices of the parallel 16-pointed rosette with a rotation of the points R, D, and E of center the point B and angle of 22.5° clockwise and counterclockwise; the points I, J, F, G, and H intersecting the extension of the sides of the petal of outer vertex R of the 8-pointed rosette and the lines r", s", r''', and s''' as shown; the perpendicular bisectors k and l of the segments of endpoints F and G, and G and H; the points K and L intersecting the lines k and r''' and l and r", respectively. Draw the part of the regular star |16/7|2 = (16, 22,5°)2 interior to the 16-pointed rosette as shown.

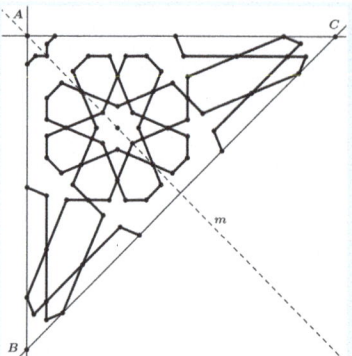

Drawing of the model 48. Step 4
4. Obtain the part of the 16-pointed rosette of the upper right side by a symmetry with respect to line m of the part of the 16-pointed rosette of the lower left side.

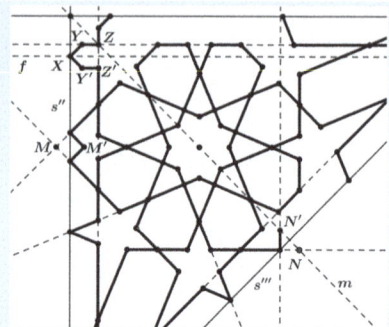

Drawing of the model 48. Step 5
5. To draw the side figures of the left side and half of the lower left side of the base unit, extend the sides of the petals of the 8-pointed and 16-pointed rosettes and compute the intersection points as indicated. Furthermore, points Y′ and Z′ are symmetric of Y and Z with respect to the horizontal line f through point X. Points M′ and N′ are symmetric of M and N with respect to the lines s″ and s‴, respectively.

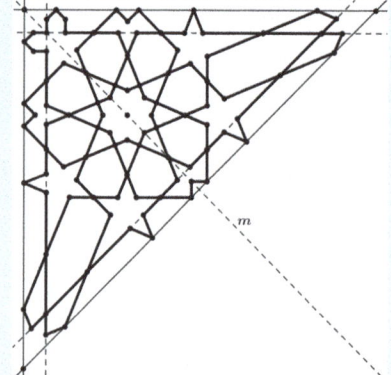

Drawing of the model 48. Step 6
6. Complete the drawing of the base unit by a symmetry with respect to the line m of the pieces drawn in 5.

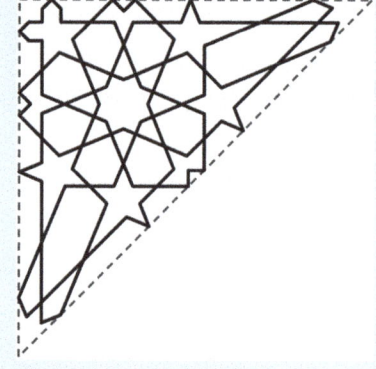

Drawing of the model 48. Step 7
7. Base unit.

4.1 Introduction

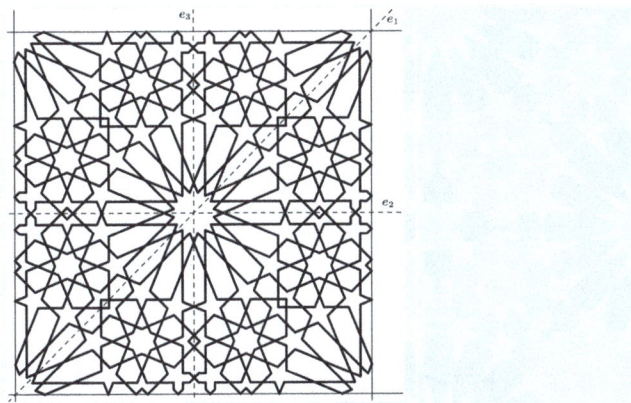

Drawing of the model 48. Step 8
8. Repeat unit obtained from the base unit by successive symmetries with respect to the lines e_1, e_2, and e_3.

Fig. 4.96 Two Sisters Room, Alhambra, Granada, Spain

Obtain the final periodic tessellation by multiple translations of the repeat unit (Fig. 4.96).

The tessellation contains parallel ideal 8-pointed rosettes of center a $(8,45°)2 = |8/3|2$ regular star, parallel 16-pointed rosettes of center a $(16,22.5°)2 = |16/7|2$ regular star, and $(8,90°) = |8/2|$ regular stars.

4.1 Introduction

4.1.49 Madrasa al-Bu'inaniya, Fes, Morocco

One can construct the tessellation of Fig. 4.97 using a repeat unit which is a parallelogram.

The design of Fig. 4.97 contains ideal parallel 8-pointed and parallel 16-pointed rosettes.

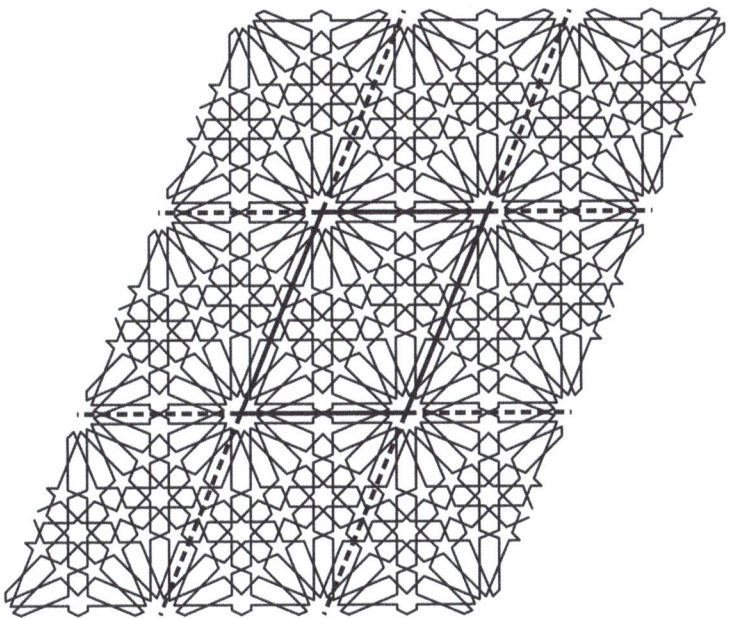

Fig. 4.97 Periodic tessellation structure of model 49

Construction of the repeat unit

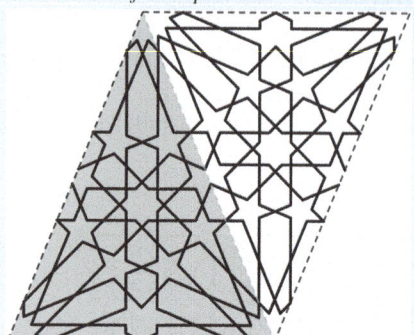

Level: Difficult

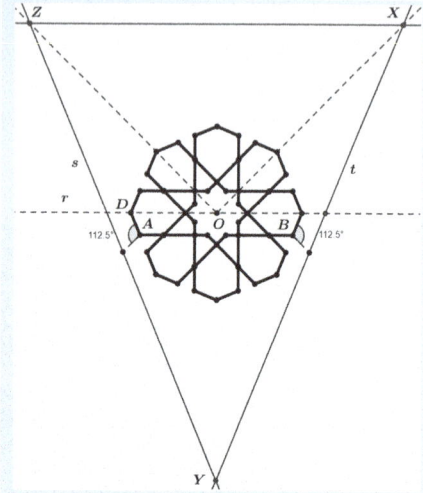

Base unit and repeat unit.
We will draw the base unit bounded by an isosceles triangle at the top right of the repeat unit in the form of a parallelogram, and then complete it with a central symmetry.
The width of the petals of the 16-pointed rosette is the same as the width of the petals of the 8-pointed rosette. The sides of the petals of the 8-pointed rosette extend to the petals of the 16-pointed rosette. Thus, the 16-pointed rosette is highly constrained.

Drawing of the model 49. Step 1
1. Start from a parallel ideal 8-pointed rosette of center a (8,45°)2 = |8/3|2 regular star. Draw: the line r through points O and D; the line s rotation of r of center the point A and 112.5° counterclockwise; the line t rotation of r of center the point B and 112.5° clockwise. Draw the points X, Y, and Z as shown. The boundary of the base unit is the isosceles triangle of vertices X, Y, and Z.

4.1 Introduction

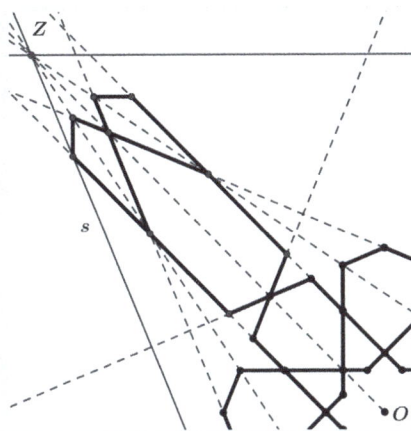

Drawing of the model 49. Step 2
2. In the upper left part of the base unit, we will draw part of a parallel 16-pointed rosette with a central (16,22.5°)2 = |16/7|2 regular star. Draw the extension of the sides of the petals of the 8-pointed rosette and the points of intersection as shown. Then complete the drawing.

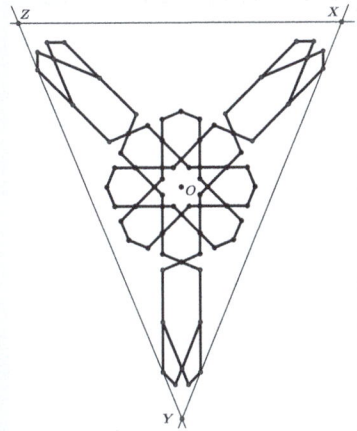

Drawing of the model 49. Step 3
3. Rotate the design of the upper left part of the base unit with a rotation of center the point O and angle of 135° counterclockwise to obtain the lower part. Rotate the lower part by a rotation of the center O and angle of 135° counterclockwise to get the upper right part of the design.

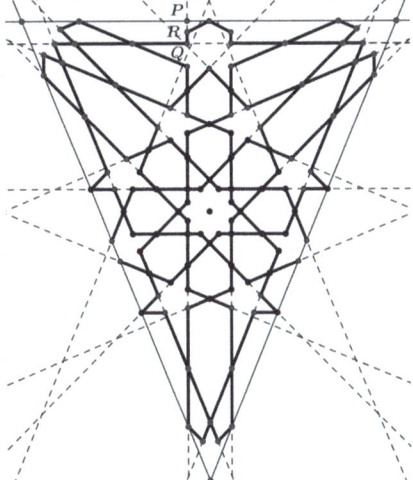

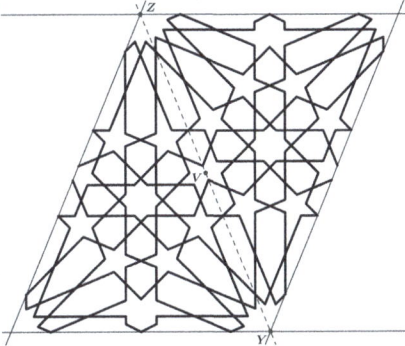

Drawing of the model 49. Step 5
5. Draw the midpoint V of the segment of endpoints of Z and Y. Obtain the bottom left part of the repeat unit by a central symmetry of center V of the base unit.

Drawing of the model 49. Step 4
4. Complete the drawing of the 16-pointed rosettes within the interstitial region of the base unit using 22.5° angle rotations centered on points X, Y, and Z. The point R is the midpoint of P and Q. Complete the drawing inside the interstitial region of the base unit as shown.

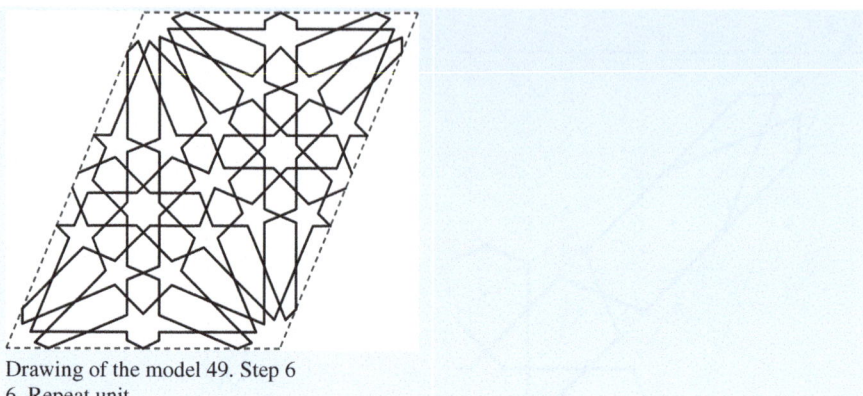

Drawing of the model 49. Step 6
6. Repeat unit.

4.1 Introduction

Fig. 4.98 Madrasa al-Bu'inaniya, Fes, Morocco

Obtain the final periodic tessellation by multiple translations of the repeat unit (Fig. 4.98).

The tessellation contains parallel ideal 8-pointed rosettes with center a $(8,45°)2 = |8/3|2$ regular star and parallel 16-pointed rosettes with center a $(16,22.5°)2 = |16/7|2$ regular star.

4.1.50 Balcony of Lindaraja, Alhambra, Granada, Spain

One can construct the tessellation of Fig. 4.99 using a repeat unit which is a rhombus.

The design of Fig. 4.99 contains parallel ideal 9-pointed and parallel regular 12-pointed rosettes.

Fig. 4.99 Periodic tessellation structure of model 50

4.1 Introduction

Construction of the repeat unit

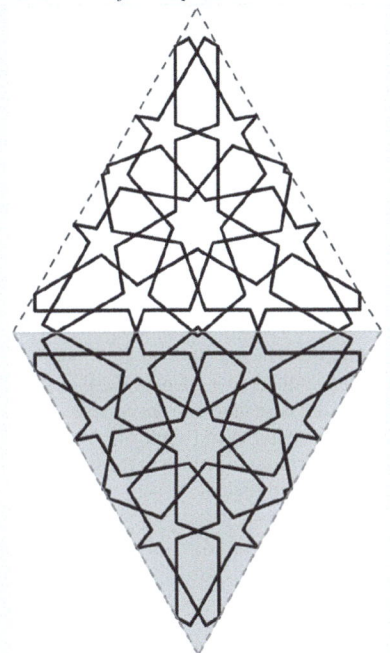

Level: Difficult

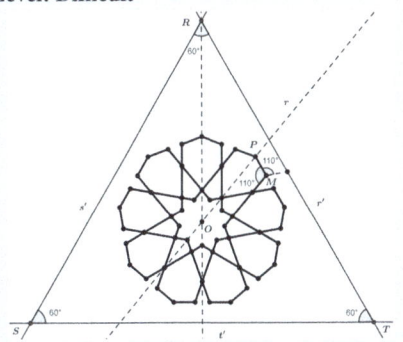

Drawing of the model 50. Step 1
1. Start from a parallel ideal rosette of center a $(9,40°)2 = |9/3.5|2$ regular star. Draw lines: r' rotation of the line r of center the point M and 110° clockwise; s' rotation of r' of center the point O and 120° clockwise; t' rotation of s' of center O and 120° counterclockwise. Draw the points: R intersection of r' and s', S intersection of s' and t', and T intersection of t' and r'. The base unit boundary f is the triangle of vertices R, S, and T.

Base unit and repeat unit
We will draw the base unit bounded by an equilateral triangle on top of the rhombic repeat unit, and then complete it with an axial symmetry.
The width of the petals of the 9-pointed rosette is the same as the width of the petals of the 12-pointed rosette. The sides of the petals of the 9-pointed rosette extend to the petals of the 12-pointed rosette. Thus the 12-pointed rosette is highly constrained.

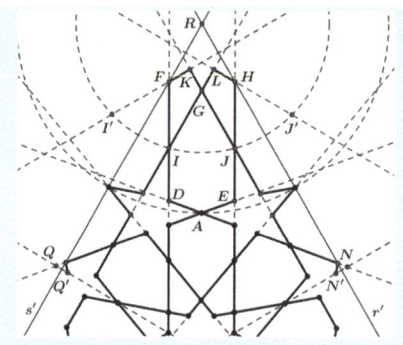

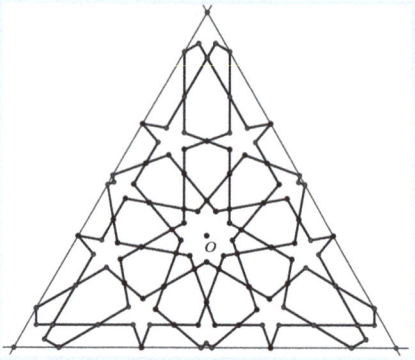

Drawing of the model 50. Step 2
2. Draw: the points D and E intersecting the extension of the sides of the petal of outer vertex A of the 9-pointed rosette; the missing vertices of the half petals of the parallel 12-point rotating points A, D, and E by rotations of center R and angle of 30° clockwise and counterclockwise; the points I, J, F, G, and H intersecting the extension of the sides of the petals of the 9-pointed rosette as shown; the points I′ and J′ rotating the points I and J by a rotation of 30° of center the point R; point K intersection of the line passing through I′ and F and the line passing through J and G; point L intersection of the line passing through J′ and H and the line passing through I and G; points N′ and Q′ symmetric of points N and Q with respect to lines r′ and s′, respectively.

Drawing of the model 50. Step 3
3. Complete the drawing within the interstitial region of the base unit with successive rotations of center O and angle 60° of the previously drawn extensions in the upper part of the interstitial region.

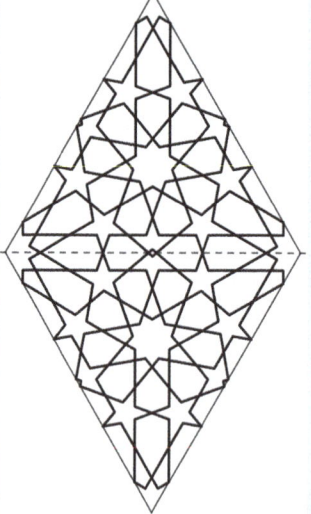

Drawing of the model 50. Step 4
4. Obtain the repeat unit from the base unit using an axial symmetry.

4.1 Introduction

Fig. 4.100 Balcony of Lindaraja, Alhambra, Granada, Spain

Obtain the final periodic tessellation by multiple translations of the repeat unit (Fig. 4.100).

The tessellation contains parallel ideal 9-pointed rosettes of center a (9,40°)2 = |9/3.5|2 regular star and parallel regular 12-pointed rosettes of center a (12,30°)2 = |12/5|2 regular star.

Annotated Bibliography and Webography

Below is a list of the books, articles, and websites I used to write this book. It is intentionally brief because, given the practical nature of the book, I think it is best to cite only the sources that really helped me prepare it.

I want to highlight that, as far as I know, there is no source on the radial extension method for generating periodic tessellations with one or more stars or rosettes as the central motif and how to use IGS to construct them described in this book.

Bonner, Jay. 2017. Islamic Geometric Patterns: Their Historical Development and Traditional Methods of Construction. Springer

The book begins with the historical antecedents, initial development, maturity, and dissemination of Islamic geometric patterns. Next, it provides a classification by design methodology and an in-depth presentation of the polygonal technique of pattern construction. The last section, contributed by C. Kaplan, explores the use of computation to generate Islamic geometric patterns. The book contains 105 photographs and more than 540 computer-generated graphics to illustrate the patterns and how they are generated.

Bourgoin, J. 2012. Arabic Geometrical Pattern and Design. Dover Publications

The book contains about 200 hand-drawn examples, in black and white, showing a wide range of Islamic geometric patterns. Dotted lines are drawn on each pattern to show squares, rectangles, triangles, hexagons, or circles that give a rough idea of the repeat unit.

Broug, Erik. 2019. Islamic Geometric Patterns. Thames & Hudson

The book provides a step-by-step guide to drawing Islamic geometric patterns using paper, pencil, compass, and ruler or alternatively using IGS. The first chapter gives details of the geometry basics, the second chapter is the step-by-step guides to draw 19 patterns classified as Easy, Intermediate, or Difficult.

Cromwell, Peter R. 2021. Looking at Islamic Patterns I: The Perception of Order. Available at: https://osf.io/preprints/psyarxiv/qhg3f, accessed June 2024

The article explores the visual properties of star patterns to identify features that contribute to a positive response in the viewer. It describes different constructions to create star motifs and investigates what makes a good star as well as develops a model to determine how the stars should be arranged to form an attractive composition.

Cromwell, Peter R. 2023. From circle packings to constellation patterns. Journal of Mathematics and the Arts. 17.1–20

The article describes the wheel construction of a star and presents the process to convert a circle packing into an arrangement of interconnected star motifs.

A. Lee, Islamic star patterns, Muqarnas 4 (1987), pp. 182–197

The article presents an introduction to the geometric logic of star and rosette patterns.

Lee, Tony and Ayman Soliman. 2014. The Geometric Rosette: Analysis of an Islamic Decorative Motif. Available at: https://tilingsearch.mit.edu/RosetteAnalysis.pdf, accessed June 2024

The article develops a geometrical analysis of both the isolated rosette motif itself and its use in a variety of repeating patterns.

Majewski, Mirosław. 2020. Practical Geometric Pattern Design: Geometric Patterns from Islamic Art. Amazon Kindle Direct Publishing

The book describes how to draw step-by-step around 50 Islamic geometric patterns with the Gereh method using paper, pencil, compass, and ruler or alternatively using IGS.

Martínez Vela, Manuel. 2022. How to draw the mosaics of the Alhambra. Editorial Almizate

The book, of 600 pages and with more than 1600 drawings and photographs, studies around 90 geometric patterns from the Alhambra of Granada. All of them are described with step-by-step instructions for constructing them with a ruler and compass by hand or using IGS. The history, diagrams, and explanations associated with the construction of each pattern are excellent.

Patterns in Islamic Art. The Wade Photo Archive. Available at: https://patterninislamicart.com/s/collections/main-archive, accessed June 2024

Photographic archive which contains over 4000 images of Islamic patterns from around the world that can be browsed in its entirety region by region. In addition to the photographic archive, there is other material related to Islamic patterns available on this website: Drawings, Diagrams & Analyses; Texts & Essays; and Published Material. There is also a Bibliography, listing a broad selection of books related to Islamic patterns.

Tiling Search Web Site. Searching Islamic patterns by location. Available at: https://tilingsearch.mit.edu/area1.htm, accessed June 2024

This database allows searching for high-quality images of Islamic patterns by location. For each pattern, the location is provided, its geometry is described, and a list of sources in which it is referenced is given.

Wichmann, Brian and Wade, David. 2018. Islamic Design: A Mathematical Approach. Birkhäuser, Springer

The book has two parts. Part I presents an overview of Islamic history, including their philosophical and scientific achievements. Part II defines the key concept of rosette; presents a formal analysis of a great range of classical patterns using mathematical techniques supported by computer-drawing software.

Further Reading

Two books are referenced below for those interested in having a deeper understanding of the nature and meaning of Islamic art.

Critchlow, K. 1976. Islamic Patterns. An Analytical and Cosmological Approach. Thames and Hudson Ltd.

The book presents a philosophical and interpretive analysis of the geometrical patterns of Islamic art, explaining their inseparability from mystical mathematics and, with the aid of nearly 200 drawings, their reflection of the cosmological laws affecting all creation.

El-Said, I. and A. Parman. 1976. Geometric Concepts in Islamic Art. World of Islam Festival Publishing Company Ltd.

The book studies abstract patterns and concepts of design used in Islamic art and architecture, employing analytical diagrams to detail the conception, design, and construction of patterns through a geometric system. It shows how abstract beauty is achieved with perfect interrelationship between the parts and the whole and irrespective of the mode, form, or scale of expression.

The manufacturer's authorised representative in the EU is Springer Nature Customer Service Centre GmbH, Europaplatz 3, 69115 Heidelberg, Germany. If you have any concerns regarding our products, please contact ProductSafety@springernature.com

Printed and bound by CPI Group (UK) Ltd, Croydon, CR0 4YY

26/03/2026

02078943-0010